1,000,000 Books

are available to read at

Forgotten Books

www.ForgottenBooks.com

Read online
Download PDF
Purchase in print

ISBN 978-1-331-55645-9
PIBN 10205448

This book is a reproduction of an important historical work. Forgotten Books uses state-of-the-art technology to digitally reconstruct the work, preserving the original format whilst repairing imperfections present in the aged copy. In rare cases, an imperfection in the original, such as a blemish or missing page, may be replicated in our edition. We do, however, repair the vast majority of imperfections successfully; any imperfections that remain are intentionally left to preserve the state of such historical works.

Forgotten Books is a registered trademark of FB &c Ltd.
Copyright © 2018 FB &c Ltd.
FB &c Ltd, Dalton House, 60 Windsor Avenue, London, SW19 2RR.
Company number 08720141. Registered in England and Wales.

For support please visit www.forgottenbooks.com

1 MONTH OF FREE READING

at

www.ForgottenBooks.com

By purchasing this book you are eligible for one month membership to ForgottenBooks.com, giving you unlimited access to our entire collection of over 1,000,000 titles via our web site and mobile apps.

To claim your free month visit:
www.forgottenbooks.com/free205448

* Offer is valid for 45 days from date of purchase. Terms and conditions apply.

English
Français
Deutsche
Italiano
Español
Português

www.forgottenbooks.com

Mythology Photography **Fiction**
Fishing Christianity **Art** Cooking
Essays Buddhism Freemasonry
Medicine **Biology** Music **Ancient Egypt** Evolution Carpentry Physics
Dance Geology **Mathematics** Fitness
Shakespeare **Folklore** Yoga Marketing
Confidence Immortality Biographies
Poetry **Psychology** Witchcraft
Electronics Chemistry History **Law**
Accounting **Philosophy** Anthropology
Alchemy Drama Quantum Mechanics
Atheism Sexual Health **Ancient History**
Entrepreneurship Languages Sport
Paleontology Needlework Islam
Metaphysics Investment Archaeology
Parenting Statistics Criminology
Motivational

A

POPULAR HISTORY

OF

BRITISH SEA-WEEDS,

COMPRISING

THEIR STRUCTURE, FRUCTIFICATION, SPECIFIC CHARACTERS,
ARRANGEMENT, AND GENERAL DISTRIBUTION,
WITH NOTICES OF SOME

OF THE

FRESH-WATER ALGÆ.

BY THE

REV. D. LANDSBOROUGH, A.L.S.,

Member of the Wernerian Society of Edinburgh, and Author of
"Excursions to the Isle of Arran."

SECOND EDITION.

LONDON:
REEVE AND BENHAM, HENRIETTA STREET, COVENT GARDEN.
1851.

PRINTED BY REEVE AND NICHOLS,
HEATHCOCK COURT, STRAND.

TO

ROBERT KAYE GREVILLE, M.D.,

EDINBURGH,

AND TO

WILLIAM HENRY HARVEY, M.D.,

PROFESSOR OF BOTANY, DUBLIN,

THIS LITTLE WORK,

IN WHICH THEY HAVE KINDLY AIDED HIM,

IS,

WITH MUCH RESPECT AND ESTEEM,

DEDICATED

BY THEIR GRATEFUL FRIEND,

THE AUTHOR.

Not lost the time in sea-side ramble spent.
Braced is the frame; and mental health is gained;
Knowledge is gained of Him who made the deep,
And blissful love acquired of nature's works,
Of which each questioned "minim" soothly says:
" The plastic power that formed us is divine!"
Launch then the skiff; ply well the scraping dredge!
Or, if it like thee better, search the shore:
Each rock-pool has its treasure, every tide
Strews on the yellow sand, from Ocean's lap,
Weeds, than our flowers more fair, and fitted more,
By Lady's gentle fingering displayed,
To beautify the album's tasteful page,
Than aught that deftest pencil e'er devised
Of graceful symmetry, or lovely hue:
For "who can paint like Nature?" quoth the Bard.—*D. L.*

PREFACE TO THE SECOND EDITION.

I HAVE been much gratified by the rapid sale of the first edition of this little work. Though I am aware that its success is not a little owing to the taste and skill of the Publisher, and to the artistic excellence of the illustrations, yet it shows, also, an increasing love, throughout our land, of the beauties of nature, and an increasing interest, I trust, in His works, who has given to nature so many instructive charms.

Though the size of this book is not greatly increased, there are in this new edition hundreds, I believe, of little additions and changes, and, I hope, improvements. I have been helped much by Professor Harvey's second edition of his excellent 'Manual,' in bringing up the history of British Sea-weeds to the present state of the science. Even during the year that has elapsed since the first edition was published, it will be found that some new plants have been added to the British lists, and that several, for the first time, have been found in Scotland, where many, I doubt not, will yet be discovered.

Without repeating my thanks to those who assisted me in the first edition, I have much pleasure in expressing my obligations to several, who, by useful information and beautiful specimens, have aided me in preparing the present edition. For most beautiful specimens of rare Algæ, I have been much indebted to the Rev. W. S. Hore, now of Norwich; to Dr. Cocks, of Devonport; and to Mr. Boswarva, of Plymouth. To Miss Cutler—a name well known to Algologists, and a lady much loved, I am convinced, by all who know her—I have pleasure in returning thanks for many specimens, and for much useful information. My kindest thanks, also, are due to Mrs. Gulson, of Exmouth, who, having gained for herself a name in Conchology, is prosecuting the study of Algology with an active intelligent zeal, which has already been crowned with much success, and which cannot fail to render her highly distinguished.

Nor am I indebted only to the kind naturalists of the sunny south : Scotia's sons and daughters have also given good aid. Lady Keith Murray has done me the honour of sending me some interesting specimens, gathered by her near Stonehaven, on the north-east coast of Scotland, and I doubt not that she will gather fresh laurels for herself on that rich though rock-bound shore. The Rev. Mr. Pollexfen kindly called on me, and gave me some beautiful Algæ from Orkney. Though now residing in London, and though his visits to the north are few and far between, a day's dredging in Lerwick Bay comes like a refreshing gleam, reminding him of the days of other years. To the Rev. Gilbert Laing I return my kind thanks for specimens from Orkney and other localities. From Mrs. C. Allardyce, of Cromarty, I have received some beautiful specimens, and with her zeal and intelligence I doubt not that she will make many interesting discoveries on that classic shore. I am under obligations, also, to Mrs. Balfour, of Edinburgh, who, when residing in the Isle of Arran last summer, with her husband, Professor of Botany in the University of Edinburgh, engaged with such zeal in Algological pursuits, that she has already added to the marine flora of Scotland, and has thus helped to save the credit of our Scottish naturalists, who, it is alleged, allow English naturalists to surpass, even when making only occasional visits to our truly-interesting land. Nor must I forget to return my best thanks to Miss Turner and to Miss White, for some of the rare and beautiful Algæ in which Jersey abounds. To many other friends I lie under obligations, which, I trust, I have acknowledged in the body of this little work.

PREFACE TO THE FIRST EDITION.

THOUGH British Algology has for several years, at leisure hours, been a favourite study of mine, I should have been afraid to undertake this introductory work had I not been encouraged by Dr. Greville of Edinburgh, and Professor Harvey of Dublin. The former lent me books, gave me advice, and allowed me to avail myself of his published works; the latter gave me counsel, allowed me to take aid from his publications, and solved my doubts respecting plants I sent him.

In describing the Corallines I was glad to avail myself of the accurate descriptions given by my excellent friend Dr. Johnston. In the fresh-water department, I was happy to draw on my friends Mr. Ralfs and Mr. Hassall. Mrs. Griffiths, with her characteristic kindness, granted me the much-valued privilege of consulting her respecting marine Algæ; and was ready at all times to favour me with specimens.

The preliminary chapters, extending to a greater length than I anticipated, will not, I trust, be regarded as altogether uninteresting. In the body of the work, I can safely recommend the description of *genera*, being chiefly taken from the valuable works of Dr. Greville and Professor Harvey. The plates are by Mr. Fitch, whose talents in this department are too well known to need praise from me. The plates add greatly to the value of the work, by making verbal descriptions much more easily understood:—

"Segnius irritant animos demissa per aurem,
Quam quæ sunt oculis subjecta fidelibus."

Though the work is intended for beginners, I shall venture to hope that some portions of it may not be quite devoid of interest to proficients in science. The *savans* of the south may not disdain to listen even to a sciolist of the north-west, when, without pretension, he records his own observations.

PREFACE.

British Algology is making as rapid progress as any other branch of natural science. Many of those persons who spend a month or two in summer on the sea-coast, have discovered that there is great beauty in sea-weeds, and have found that there is great pleasure in preparing little collections of these marine paintings, to gratify their inland friends on their return, and to afford to themselves pleasant reminiscences of happy hours spent in healthful recreation on the sea-shore; and often, we trust, in devout musings on Him whose path is in the deep waters; whose hand has formed all the wonderful works they contain, and whose voice may be heard in the gentlest whisperings of the waves, or in the mighty noise of the great sea-billows.

> "The gentleness of Heaven is on the Sea:
> Listen! the mighty Being is awake,
> And doth with his eternal motion make
> A sound like thunder—everlastingly."—*Wordsworth.*

Rockvale, Saltcoats,
   May, 1849.

LIST OF PLATES.

Plate I.

Fig.		Page.
1	Fucus serratus	110
2	Halidrys siliquosa	104
3	Desmarestia ligulata	129
4	Alaria esculenta	116

Plate II.

5	Dictyota dichotoma	141
6	Cutleri multifida	136
7	Asperococcus Turneri	148
8	Striaria attenuata	144

Plate III.

9	Chorda filum	151
10	Haliseris polypodioides	137
11	Myriotrichia clavæformis	164
12	Sargassum bacciferum	96

Plate IV.

Fig.		Page.
13	Himanthalia lorea	114
14	Cladostephus verticillatus	154
15	Sphacelaria scoparia	156
16	Punctaria latifolia	145

Plate V.

17	Sporochnus pedunculatus	132
18	Asperococcus compressus	147
19	Ectocarpus siliculosus	159
20	Mesogloia virescens	170

Plate VI.

21	Callithamnion gracillimum	182
22	—— floccosum	178
23	—— corymbosum	183
24	Seirospora Griffithsiana	187

Plate VII.

Fig.		Page.
25	Callithamnion tetricum ..	177
26	——— roseum	178
27	Ceramium diaphanum ...	197
28	Ptilota sericea	205

Plate VIII.

29	Ceramium rubrum	197
30	Ptilota plumosa	204
31	Chylocladia ovalis	273
32	Microcladia glandulosa...	203

Plate IX.

33	Chondrus crispus	221
34	Phyllophora rubens	223
35	Iridæa edulis	216
36	Rhodymenia laciniata....	243

Plate X.

37	Hypnea purpurascens	237
38	Ceramium acanthonotum .	198
39	Catenella Opuntia	217
40	Dumontia filiformis	229

Plate XI.

41	Bostrychia scorpioides ...	288
42	Gigartina acicularis	231
43	Rhodymenia jubata	245
44	Gracilaria confervoides ...	239

Plate XII.

Fig.		Page.
45	Bonnemaisonia asparagoides	266
46	Plocamium coccineum ...	249
47	Polysiphonia parasitica .	291
48	Sphærococcus coronopifolius	241

Plate XIII.

49	Delesseria ruscifolia	260
50	——— sanguinea	252
51	Nitophyllum punctatum ..	261
52	Halymenia ligulata......	227

Plate XIV.

53	Chylocladia articulata....	274
54	Corallina officinalis......	276
55	Polysiphonia elongella ...	301
56	Chrysimenia clavellosa ...	271

Plate XV.

57	Chylocladia kaliformis ...	273
58	Laurencia pinnatifida	268
59	Rytiphlæa thuyoides	290
60	Polysiphonia Brodiæi	296

Plate XVI.

61	Ulva Linza	330
62	Enteromorpha compressa .	328
63	Codium tomentosum	307
64	Bryopsis plumosum	309

Plate XVII.

Fig.		Page.
65	Cladophora lanosa	314
66	—— arcta	315
67	—— rupestris	313
68	—— rectangularis	313

Plate XVIII.

69	Conferva tortuosa	318
70	—— Melagonium	318
71	Enteromorpha erecta	328
72	—— intestinalis	326

Plate XIX.

Fig.		Page.
73	Ulva latissima	329
74	Padina pavonia.........	138
75	Porphyra laciniata	320
76	Laminaria Phyllitis	125

Plate XX.

77	Batrachosperm. moniliforme	350
78	Cladophora glomerata....	344
79	Draparnaldia nana	357
80	Micrasterias denticulata ..	364

PLATES OF FRUCTIFICATION.

(*Figures more or less magnified.*)

A.—Polysiphonia urceolata.
    Fig. 1. Portion of a branch.
        2. Capsule more highly magnified.

B.—Polysiphonia urceolata.
    Fig. 1. Portion of a branchlet, showing granular fruit (tetraspores).
        2. Tetraspore, highly magnified.

C.—Ptilota plumosa.
    Fig. 1. Involucre containing favellæ.
        2. Lacinia with tetraspores on short pedicels.

D.—Plocamium coccineum.
    Fig. 1. Branchlet with a lateral tubercle containing spores.
        2. Branchlet with stichidia containing tetraspores.

E.—Odonthalia dentata.
    Fig. 1. Capsular fruit.
        2. Stichidia, containing tetraspores.
           (Observed in abundance in early winter.)

F.—Rhodymenia bifida.
   Fig. 1. Portion of frond with imbedded tubercles (coccidia), containing spores.

Nitophyllum punctatum.
   Fig. 2. Portion of frond exhibiting one of the sori, consisting of an assemblage of tetraspores.

G.—Callithamnion Borreri.
   Fig. 1. Branch with capsular fruit
   2. Branchlet with a bilobed

H.—Phyllophora rubens.
   Fig. 1. Wart, with one leaf rem show the nemathecia.
   2. Moniliform filaments, co the nemathecium.

INTRODUCTION.

THE time is not very far gone past, when a book on the study of Sea-weeds would have been the very reverse of popular in any case. About fifty years ago, in some academic chairs, they were treated with disdain. We have heard of a student about that period who, having collected some beautiful Algæ on the shore, showed the contents of his vasculum to the Professor of Botany whose lectures he attended, expressing a wish to get some information respecting them. The Professor looked at them, and putting on his spectacles, again looked at them, when, pushing them from him, he exclaimed: "Pooh! a parcel of Seaweeds, Sir; a parcel of Sea-weeds!" The Newhaven fishermen seem to have caught the spirit of this learned Professor, for to this day do they denominate all the finer Sea-weeds,—*chaff*. They are *weeds* : and what are weeds? Dr. Johnson, the famous lexicographer, tells us that they are plants that are *noxious* and *useless*. Oh, but they are *sea*-weeds, we say in mitigation. And does that mend the matter? Horace and Virgil knew that right well, and as with all their poetic genius they knew not a word of English, they gave them the outlandish name of *Algæ* :—nay more, Horace speaks of them as "*inutilis Alga,*" useless Seaweeds! And tasteful Virgil goes even beyond his friend Horace, for, when speaking of something which he regards as worthless and filthy, he says that it is "*Algá projectá vilior,*" viler than the Sea-weed cast out on the shore. Its very calamities are turned against it: "*Refunditur Alga,*" says another poet, the sea loathes it and flings it out on the land!—Alas for the poor Sea-weeds, when the princes of poetry in the Augustan age are against them! But, as there were no Sea-weeds in the streets of Rome or in the yellow Tiber, they may have spoken thoughtlessly, and without any malice prepense, and we would therefore appeal from the ancient court of the Muses, and consider what character they sustain in the present day.

In Scotland, Sea-weeds go very generally under the name of *wrack*, or in the south and west of Scotland *wreck*, and not unfrequently *vreck*; and in this we have one of the numberless instances of the effect which the great intercourse in ancient times betwixt France and Scotland had on our Scottish language;—for what is *wrack* or *wreck* or *vreck*, but the French word *varec*, which signifies *sea-weed*. *Vraic* is the word employed in the Channel Islands, in Guernsey and Jersey, and is evidently of French extraction; and they who read Inglis on the Channel Islands will find that *vraic* is not a thing to be spoken of contemptuously! But more of this anon.

In the list of names we may mention that they often go under the name of *sea-ware*; hence we have *kelp-ware*; and even *hen-ware*, and *honey-ware*. Among the learned, *Plantæ marinæ, marine plants*, is often employed; but this term is too comprehensive, for it would include *Zostera marina*, sea-ribbon, which is a flowering plant. *Thalassiophytes*, a still more learned name, has often been given them, derived from two Greek words signifying sea-plants.

As many of them, at one time, were ranged under the Linnean genus *Fucus*, we have learned volumes descriptive of numerous *genera* and *species* then included in that comprehensive *genus*.

Fucus is the Latin form of a Greek name given to sea-weeds, and as the word signifies paint, it may have been applied to them because of the fine colour which some of them yield.

CHAPTER I.

HISTORY OF ALGÆ.

" Non ego te meis
Chartis inornatum silebo,
Totve tuos patiar labores
Impune———carpere lividas
Obliviones.'—*Horace.*

We shall mention very briefly some of those authors whose works have greatly helped to promote this department of botanical science, and with one exception we shall confine ourselves to writers of our own country. The justly-celebrated Linnæus is the exception. His attention was chiefly directed to phænogamous or flowering plants. His situation, at a distance from the sea, was not favourable for the investigation of Marine Algæ; his herbarium contained comparatively few species. His division of this order of plants was a simple one, for he ranked all aquatic Algæ, whether from the sea or from fresh water, under four great genera, *Tremella, Fucus, Ulva,* and *Conferva.* In our own country we can boast of Ray, that great and good man, whose works were of so much service in almost every branch of science, and who enumerated a good many Algæ in his Synopsis. Dillenius deserves to be mentioned, whose figures of many of the *Confervæ* are good. Hudson is worthy of much praise, and I have a regard for him because his 'Flora Anglica,' which is remarkable for its accuracy, was the first botanical work I ever saw, with the exception of Lee's Introduction. One of the next I fell in with was Lightfoot's 'Flora Scotica,' a very interesting book, and the more valuable as many of the species he describes were gathered and figured by himself. Stackhouse's 'Nereis Britannica,' which appeared in numbers at the end of the last and beginning of the present century, though now rather rare, is a work of considerable merit. For this and other valuable botanical works, I have been indebted to Dr. Robert Kaye Greville. The publication of Dr. Greville's own works at a later period, formed an era in the history of Marine Botany. They are exceedingly good, and have been highly lauded by all subsequent writers on this department of Science. His descriptions are admirable, and his exquisite plates in his 'Algæ Britannicæ,' his 'Scottish Cryptogamic Flora,' and his other publications, were well fitted to lay the foundation of his recently-acquired fame as a first-rate landscape painter.

Previous, however, to the publication of Dr. Greville's works, there were other publications that ought not to be passed over. An excellent account of the method of Dr. Walker, the well-known Professor of Natural History in the University of Edinburgh, is given by my worthy, kind-hearted friend, Dr. Patrick Neill, in his very able article in the 'Edinburgh Encyclopædia,' on *Fuci*, which has received the most unquestionable praise by being largely quoted by all subsequent writers on the subject, to which, as a rich quarry, I have already had recourse, and to which I am sure the author will make me welcome again and again to return.

Of another voluminous work, at present lying before me, it would not be easy to speak in terms of too high approval, we mean Mr. Dawson Turner's 'Historia Fucorum,' illustrated by coloured engravings chiefly by Mr. (now Sir) William Jackson Hooker, who has, by his pencil and by his pen, not only sustained but greatly increased his fame thus early acquired. Respecting the descriptions and the illustrations, Sir James Edward Smith remarks, "Never was there a more perfect combination of the skill of the painter and the botanist, than in this work."

Nor must Miss Hutchins be omitted, whose name will long be honoured by Algologists. Mr. Dawson Turner, who was so well able to appreciate her worth, pays a beautiful and tender tribute to her memory. " But few, if any besides myself, can appreciate her many amiable qualities; her liberality, her pleasure in communicating knowledge, her delight in being useful, the rapture she felt in tracing the works of the Divine Hand, and the union in her of those virtues which embellish and improve mankind. Three years have now elapsed since she died, and every succeeding year makes me feel more deeply what I have lost, and how with her is gone a great part of the pleasure I derived from these pursuits.

> " In every season of the beauteous year
> Her eye was open, and with studious love
> Read the Divine Creator in his works.
> Chiefly in thee, sweet spring, when every nook
> Some latent beauty to her wakeful search
> Presented, some sweet flower, some virtual plant.
> In every native of the hill and vale
> She found attraction, and, when beauty fail'd,
> Applauded odour or commended use.
>
> " Heu! quanto minus est
> Cum reliquis versari
> Quàm tui meminisse ! " *

Mr. Dillwyn's 'History of British Confervæ' cannot but be prized by all who are acquainted with it, as it has so

* From Mr. D. Turner's 'Historia Fucorum.'

much to recommend it, both in the plates and descriptions. Great also were the services rendered to Algology by the late Captain Carmichael of Appin, the value of whose MSS. has been amply acknowledged by Sir W. Jackson Hooker, into whose possession they came. There are several others deserving of great praise, that we are constrained to pass over; but there are two whom it would be altogether unjustifiable to omit in this limited list. The one is a lady who, so far as we know, has published nothing in her own name,—but who yet may be said to have published much, as she has so often been consulted by distinguished naturalists, who have been proud to acknowledge the benefit they have derived from her scientific eye and sound judgment. We mean Mrs. Griffiths of Torquay, of whom we may speak as *facile Regina*, the willingly acknowledged Queen of Algologists. The other to whom I refer is Professor W. H. Harvey of Dublin, who is well known by his works, and to us best by his excellent 'Manual of British Algæ,' which since its publication has been our favourite vade-mecum, till it has in some measure been supplanted by a still greater favourite, his splendid 'Phycologia Britannica,' in progress of publication, and which comes to us as a monthly feast, with its accurate descriptions and magnificent illustrations.

CHAPTER II.

STRUCTURE AND SUBSTANCE.

"He that enlarges his curiosity after the works of Nature, demonstrably multiplies the inlets to happiness; and, therefore, the younger part of my readers, to whom I dedicate this vernal speculation, must excuse me for calling upon them to make use at once of the spring of the year and the spring of life, and to acquire, while their minds may be impressed with new images, a love of innocent pleasures, and an ardour for useful knowledge; and to rememember, that a blighted spring makes a barren year, and that the vernal flowers, however beautiful and gay, are only intended by Nature as preparatives to autumnal fruits."—*Dr. Johnson.*

ALGÆ form part of that great class to which Linnæus has given the name of Cryptogamia, because they are flowerless; but, like ferns and mosses, and other plants of the same great class, they have what answers the purpose of flowers. In Lindley's valuable work 'The Vegetable Kingdom,' they form a portion of his first class in his natural system, which he denominates *Thallogens,* and they constitute the first order of this class, under the name of *Algales,* which he describes as " cellular flowerless plants, nourished through their whole surface by the medium in which they vegetate; living in water, or very damp places; propagated by zoospores, coloured spores, or tetraspores." That portion of *Algales,* or Algæ, of which we mean more particularly to treat, are called Sea-weeds, and, as their name implies, either grow in the sea, or in places where they are, occasionally at least, covered by the tide, washed by the waves, or moistened by the spray. Though very different from each other in form, colour, and general appearance, they are all similar, in so far as they are composed of cellular tissue. They have not continuous vessels like the phænogamous plants, but consist of cells differently arranged, or of gelatine, membrane, and endochrome. They have no woody fibre, though there is certainly the approach to it in appearance, as in the stems of the Great Tangle. Lamouroux, indeed, thinks that the stem of *Laminaria digitata* is formed of four distinct parts, analogous in situation, size, and organization to the *epidermis, bark, wood,* and *pith* of dicotyledonous plants. That Marine Algæ are not furnished with continuous vessels like land-plants, is obvious from the well-known fact, that if one part of a sea-plant is plunged in water, and the remainder is exposed to the air, only what is in the water remains fresh, the rest withers and becomes dry. And in the same manner, if a dried specimen of Sea-weed is in part immersed in water, while the part that is immersed becomes filled with the fluid and assumes a fresh appearance, the part that is not in the water remains dry as before.

The *fronds* of Algæ are not only variable in form, but also in substance. Some are like masses of jelly, such as the fresh-water *Batrachospermum* and *Draparnaldia*; others are very gelatinous, such as the marine *Mesogloia*; others are like silk threads, others are so filmy and membranaceous that by children they are called Sea-silk; others are cartilaginous as gristle and elastic as India-rubber, a quality which I particularly observed in a *Micromega* that I found in Arran; some are tough as leather, others firm as wood. The leaves of some are delicate and transparent, while those of others are thick and opake; some are finely veined, others are without nerves; some of the filamentous kinds have joints and longitudinal siphons; some are destitute of both. The *Diatomaceæ*, both fresh-water and marine, have an organized covering of flint, which withstands the action of fire, so that they are found abundantly in a fossil state in the substance called *Bergmehl*, or mountain-meal, extensive strata of which occur on the continent; and the microscope detects them also in a portion of Bergmehl in my possession, from a stratum of it lately discovered in Rasay, one of the islands of the Hebrides.

No part of the structure of Sea-weeds has more universally attracted attention, than the inflated portions of the stem or frond resembling bladders. These are seen in many native species, and they are very conspicuous in *Fucus vesiculosus* and *Fucus nodosus*. They are called *vesicles*, and sometimes *air-vessels*: those on *F. nodosus* are very remarkable, and escape not the attention of children, who value them the more, because they enable them to play off a practical joke on their inland friends who visit the shore, and whom they delight to startle by the explosion which the heated air occasions, when the vesicles are cast into the fire.

It is generally supposed that the vesicles are intended to give buoyancy to sea-plants, and the wisdom of God is beautifully manifested in making them at once useful and ornamental. It has been stated that the plants that are furnished with them, cease to float when the vesicles are cut off. Mr. Darwin, that remarkably scientific observer and sound reasoner, mentions, in his most interesting Journal of the Voyages of the Beagle, some Algæ that grow on the rocks in the Arctic Seas, which, though of prodigious length, instead of being spread along the bottom, as otherwise they would be, are in part floated on the surface by means of the numerous air-vessels they contain. A portion of one of these gigantic sea-plants, said to be at times 1,500 feet in length, is now in my possession. A section of it is so full of vesicles that it looks like a honeycomb, or like a piece of wood perforated by that indefatigable borer, *Xylophaga dorsalis.*

CHAPTER III.

COLOUR.

> " Not a flower
> But shows some touch in freckle, streak, or stain,
> Of *His* unrivalled pencil. He inspires
> Their balmy odours, and imparts their hues,
> And bathes their eyes with nectar, and includes
> In grains, as countless as the sea-side sands,
> The forms with which He sprinkles all the earth.
> Happy who walks with Him! Whom what he finds
> Of flavour, or of scent, in fruit or flower,
> Or what he views of beautiful or grand
> In Nature—from the broad majestic oak
> To the green blade that twinkles in the sun,
> Prompts with remembrance of a present God.
> His presence, who made all so fair, perceived
> Makes all still fairer."—*Cowper.*

THE goodness of God is remarkably manifest in the variety of sweet and beautiful colours which he has imparted to his works. This is observable in the prismatic colours of the covenant-bow, in the purple streakings of the sky, in the orient tints of the morning, and in the splendid drapery of evening clouds, when the sun seems retiring to his slumbers in the west. It is seen, also, in the green pastures, in the rosy flowers, in the verdant foliage of the groves in spring, and in the rich and mellow colouring of the woods in autumn. The Almighty, no doubt, could have preserved us in life, though all nature had worn a dull monotonous aspect. But mark the surpassing kindness of Him,

" Who, not content
With every food of life to nourish man,
Hath made all Nature beauty to his eye
And music to his ear."

" The works of the Lord are great, sought out of all them that have pleasure therein;" and, as if to encourage us to *seek them out*, and to trace Him, whose way is in the sea, and whose path is in the great waters, and whose footsteps are but little known, He shows us, by the lovely colouring which He imparts to the denizens of the deep, that even the deep feels his benignant presence;—that if we could take the wings of the morning and dwell in the uttermost parts of the sea, there would his hand lead us and his right hand hold us, and the tiniest plant that grows in the abyss would say to him who in humility seeks it out, " Behold the wonders of thy God."

To speak after the manner of men, the Most High seems so set on recommending to our notice the plants of the sea by their great beauty, that He issues, so to speak, a new law of nature. Looking at land-plants, we would say that it is a law of nature (that is, the appointment of God) that the full light of the sun is necessary to bring forth their beautiful tints. Shut out almost all the light of the sun from a rose-bud coming into flower, and, if it expand at all, how pale and sickly does it look,—

"Like saddest portrait, painted after death."

Exclude altogether the enlivening sunbeams from plants of the sweetest verdure, and they become quite blanched, as if white robes were with them, as among some nations, the weeds of woe. How different is the case with marine plants! Humboldt mentions a *Fucus* of a fine grass-green colour, brought up from the great depth of 192 feet, where it had vegetated, though the light that reached it at that depth could have been equal only to half the light of an ordinary candle; and according to his own experiments, common garden-cresses exposed during vegetation to the brilliant light of two Argand lamps, acquired only a slight tint of green. Sea-plants of a red colour, it is well known, acquire their richest red in the deepest water; and at depths to which, it is known, the rays of the sun do not reach, there are many species of Algæ of different hues, as fully coloured as those that come under the full influence of solar light.

Thus, in a way that we cannot explain, does the Lord work wonders in the deep, adorning his handiworks.

> " For not to *use* alone did Providence
> Abound, but large example give to man
> Of grace, and ornament, and splendour rich,
> Suited abundantly to every taste,
> In bird, beast, fish, winged and creeping thing,
> In herb and flower."

The prevailing colours that sea-weeds exhibit are green, olive, and red, in all their variety of shades. Those of a green colour generally grow in shallow water, the olive in deeper, and the red in deeper still; but there are many exceptions, for while the darkest olive plants are at times found in very deep water, I have seen in shallow pools *Sphacelaria plumosa*, for instance, of the darkest olive that I have ever observed in any Alga in a fresh state. And while plants of the richest red or purple are at times brought up from depths profound, *Bangia fusco-purpurea* becomes of the richest dark-purple on rocks facing the sun, and but a little within high-water mark. Brownish-olive and greenish-olive coloured plants are generally found about half-tide level. Those again that in deep water are red or purple, such as *Ceramium rubrum* and *Laurencia pinnatifida,* lose their fine colour when growing near high-tide mark, the former becoming a dirty white or yellow, and the latter a kind of olive-green. The same may be said of *Chondrus crispus,* which is purple in deep water, and green when growing in shallow pools. I have often observed the iridescence of this plant. This iridescence, I understand, is still more remarkable in *Cystoseira ericoides,* but it is rarely found on our western shores.

Having spoken of the *laws of nature,* allow me to remind my young friends of the great danger of putting *Nature* in the place of God, and of forgetting that the God of nature is also the God of grace. It would not avail us that we admired his manifold works of nature, if we were not interested in the greatest and most wonderful of all his works—the work of Redemption through his Son, Jesus Christ.

> "The Lord of all, Himself through all diffused,
> Sustains, and is the life of all that lives.
> Nature is but a name for an effect,
> Whose cause is God. He feeds the sacred fire
> By which the mighty process is maintained,
> Who sleeps not,—is not weary; in whose designs
> No flaw deforms, no difficulty thwarts,
> And whose beneficence no change exhausts."

CHAPTER IV.

VEGETATION OF SEA-WEEDS.

> Acquaint thyself with Nature. Nature's God
> Exclude not from his works. Know and adore,
> Yea, love Him as a Father;—with Him walk.
> Then, sweet to roam and trace the tiny brook
> To where it bubbles from its parent fount,
> And mark, as it meanders through the vale,
> How the rath primrose smiles on sunny brae,
> And drooping hyacinths perfume the dell—
> Sweet, too, to climb the mountain's heath-clad brow,
> To cull, 'mid cliffs, the haunt of ptarmigan,
> Rare alpine flowers that scorn the lowly vale.—
> But sweeter far to float upon the deep
> And gaze with wistful eye on all below,—
> On groves of olive tangle, intertwined
> With bright festoons of gayer, gentler algues,
> Subundane drapery, so rich and fair,
> That, like the pearl-diver, one is prone
> To cleave with downward plunge the sea-green wave,
> To grasp, and bear aloft the tempting prize,
> As trophy gained from mermaid's gay parterre.—D. L.

LAND-PLANTS are divided into annual, biennial, and perennial; and this seems to be the case also with marine plants. Many of the more tender kinds are evidently annual, nay, some of those that appear early in spring, have answered the purposes for which they were created, and have passed away before summer is far advanced. Others, though of a delicate fabric, are more enduring. They outlive the summer, and though they die partly down on the approach of winter, they send out, when spring returns, fresh fronds from the old stumps; the old and the new, though conjoined, retaining a marked difference of appearance during the early part of the season. Those of a woody fabric, like the Great Tangle, often bear considerable evidence of having weathered several winters. The mode of growth in *Laminaria saccharina*, the Sweet Tangle, and in *L. digitata* is very remarkable. The new growth begins at the base, and pushes the old portion before it. This strikes us as curious, and yet it should be familiar to us, for it is the way in which the nails of our fingers and toes are renewed.

Very interesting information respecting the rapidity of the growth of some of the large sea-weeds is recorded in Dr. P. Neill's article on *Fuci*, from which, like my predecessor, I am glad to borrow the facts observed in the course of the arduous undertaking of erecting a beacon on the Carr Rock in the Frith of Forth. The observer was that highly respectable civil-engineer, Mr. Stephenson, and the observations were made at the request of his friend Dr. Neill, to whom specimens of the sea-weeds were transmitted. The Carr Rock is at the entrance of the Frith. It is about twenty feet broad and sixty feet long, and it is uncovered only at the lowest ebb of spring tide. When the operations were begun, it was clothed with large sea-weeds, especially with the Great Tangle, *Laminaria digitata*, and *Alaria esculenta*, or Badderlocks. In the course of 1813 the workmen succeeded in clearing and levelling a considerable portion of the foundation of the intended building, but in the beginning of November operations were necessarily abandoned for the winter. At this time the rocks by pick and axe had been made quite bare; the sea-weeds had been cut away, the roots trampled, and much of the rocks had been chiselled, so that the very stumps had been cleared away. On returning to the rock in May 1814 to resume operations, it was matter of no small surprise to find the rock as completely covered with large sea-weeds as when they first landed on it, though little more than six months had elapsed since they left it quite bare. In particular, it was observed that many recently produced specimens of *Alaria esculenta* (Badderlocks) measured six feet in length, which is above its average length at full size, and they were furnished with the pinnated appendages at the base which contain the seed of the plant. The specimens of common Tangle were generally only about two feet in length, whereas when fully grown they may be four feet and upwards. The specimens of both these Algæ were taken from that part of the rock which had been dressed with the pick and chisel, before the workmen left it in autumn, so that they had evidently grown from the seed; and indeed it was observed that the seaweeds had grown more luxuriantly on the recently dressed sandstone rock, than on those parts where the stumps had been only trodden down. It appears, therefore, that the seeds floating in the waves must have attached themselves to the rock after the middle of November, and must have vegetated and increased thus rapidly during a winter which many are old enough to remember was one of great severity.

It is wonderful that so few have attempted the growth of sea-weeds from seed, especially as Mr. Stackhouse more than fifty years ago set the example, and showed that it is quite practicable. The account of his experiments, were it not rather too long, I would willingly extract from his 'Nereis Britannica' now before me. The substance of part of it is as follows:—He got wide-mouthed jars with a siphon to draw out the water without shaking it. On the

7th of September, 1796, he placed plants of *Fucus canaliculatus* in the jar with their bases downwards. On the following morning he decanted the water into a basin. He then poured a fresh quantity of sea-water on the plants, and placed the jar in a window facing the south. On the following morning the plants discharged a few yellowish grains, which proved to be the actual seed of the plant;— these seeds, however, were not in contact with the water, but each enveloped with a bright mucilaginous substance, which, from being heavier than the water, made it sink, and caused it to adhere to the rock. Watching these seeds, he had the pleasure of seeing one of them exploding so as to agitate the water, from which he learned that some sea-weeds, when ripe, scatter their sporules by the bursting of the capsules without waiting for the decay of the frond.

He next got pebbles from the sea-beach, and having drained off the greater part of the water, he poured the remainder on the pebbles and left them to dry for some time that the seeds might adhere to them. He then fastened strings to the pebbles, and alternately sank them in the jar, and drew them out and left them exposed to sun and rain, in imitation of what would have been experienced by them by ebb and flow at mid-tide mark, had they been in their natural situation on the shore. In less than a week a thin film was discovered on the surface of the pebbles. It gradually increased in breadth and thickness, until at last he observed buds arising from the membrane. These central shoots increased in size, but not rapidly after the first efforts; and as he had not an opportunity of placing them in a rock-pool, owing to his being miles from the shore, he discontinued the experiments.

More recently Algæ have been raised from seed by J. A. Agardh and by some naturalists in our own country, but still much remains to be done. Even young Algologists might make experiments in this department. I confess that I have done little in this way myself, and the little that I have done has been conducted in such a manner as to lead to no practical results. Yet it has not been without interest, and I shall therefore mention it as an encouragement to my young friends, who may have more time on their hands and more dexterity.

In the very end of September, 1848, D. Landsborough, jun., had brought from the sea some rare Nudibranchs, which he put in a tumbler of sea-water and placed in a window with a south-east exposure. They lived there for several weeks, and when they began to look feeble, they were returned to the sea as a reward for their good behaviour. Before I granted manumission to the beautiful Nudibranchs, I had observed at the bottom and on the sides of the tumbler the growth of young Algæ. The first that I observed were grass-green, consisting of simple filaments without any visible joints. Many of these at the end of five months are still alive, but during the cold winter months they have not increased in size, and though they are visible by the naked eye, they are not above a line in length. There are one or two of the same size of a reddish-purple colour. There were also a number of little dense tufts of a brownish-olive colour, the crowded filaments of which were like those of *Sphacelaria*, but terminating in white hairs longer than the filaments, but so fine as to be invisible to the naked eye. They answered in some degree the description in Harvey's Manual of *Chætophora Berkeleyi*; yet if the name means *bristle-bearing*, it would not correspond, for the fine white tippings of the filaments are too flexible and waving to be called *setaceous*. In the body of the water there were a few long filaments almost colourless, finer than human hair, and so limber that they bent under the weight of the almost invisible infusoria, when they rested from their sportive gambols. Then there were others that were just perceptible, as small dots, by the naked eye, but when seen through a pretty powerful lens they were perfectly circular and of beautiful workmanship, and not unlike some of the figures of *Desmidieæ*. These in general were green, but a few were reddish-brown; and last of all there were a number of very minute branched Algæ, just perceptible as a faint haze by the naked eye, but when examined by a lens they were more like exceedingly diminutive specimens of *Sphacelara filicina* than anything I remember, though I do not at all think that they were the young of this plant. The branches adhere closely all along to the glass, spreading in all directions from a centre, though not with very uniform regularity. The more advanced branches seem to lose their pinnæ towards the point; they, and most of the others, are still alive in the tumbler, which would form a studio not unworthy even of Mr. Ralfs. The difficulty with me, besides want of time, is how to remove such exceedingly minute and delicate structures so as to bring them under the field of the microscope. Much do I wish that they were in the hands of a good microscopist, to examine them, and still more minute *Vorticellæ* which in beautiful tufts are mingled along with them, but so small that a tuft of thirty is not perceptible, even as a haze, by the naked eye. I may mention that once a fortnight I pour off the water and give a fresh supply from the sea.*

* As my object is to aid in rendering my young friends not merely Algologists, but diligent observers of the phenomena of nature, I shall not consider myself bound to adhere rigidly to one department of nature's works. To encourage them in their researches, I may mention that a single tumbler of water will furnish a rich field for their bright young eyes. This very tumbler which showed me the germination of Algæ from seed, and which exhibited also the beautiful *Vorticellæ*, contained numberless *infusoria* of many kinds, merrily dancing in all directions, and showing that he who made them, blessed them with happiness. These animalculites I had seen before, but in watching their sportive gyrations, I was gratified with appearances that I had never before observed. Perceiving what I thought a little hazy spot on the glass, I applied a lens, and found that it did not adhere to the glass, but was moving up and down. Afterwards more than a score were observed, some of them little semi-pellucid, and, I think, hollow balls;—others more like broad flattened bonnets, such as are worn at times by carriers;—with an aperture for the reception of the head. The largest, however, were less than a line in diameter, and of a light grey colour. When the tumbler was allowed to remain unmoved, they lay invisible at the bottom; but when it was gently agitated, they mounted up like little balloons to the surface of the water, and then gradually descended. How they moved I could not tell. The surface of the balls in certain lights seemed a little hirsute, but I could observe nothing like the motion of cilia. When they were all in motion, some ascending and others descending, the mystic movements of these little spheres presented a very animated spectacle.

But what were my little peripatetic puff-balls? At first I despaired of being able to tell; but fortunately I had beside me Sir J. G. Dalyell's recent publication, and turning over its pages and plates I was delighted to find that what I had contemplated with so much interest was the *progeny of Medusa*,

As Sea-weeds grow,—how and whence do they obtain nourishment? Land-plants, it is known, derive nourish-

for in his plate xxi. his figures quite corresponded with what I had observed. I then tried an experiment on them which Sir John does not mention having done. I took the tumbler into a darkened apartment, and giving the glass a smart percussion, instantly my little puff-balls sent forth a very brilliant flash of phosphorescent light, showing me that in all likelihood they play no very secondary part in that beautiful phosphorescence of the sea, which in the wake of a vessel I had so often admired in a summer evening.—I continued to watch them in the hope of seeing them transformed into *Medusa bifida*, but frost of unusual intensity for the season set in after the middle of October, and my *Medusettes* sank under it. On trying to rouse them, only one attempted to rise, and next day it had vanished,—like another creature of greater pretension, "fleeing also as a shadow and continuing not."—On contemplating the wonderful works of God even in this little world of water, one is led to exclaim, in the singularly beautiful and truly eloquent words of Hedwig:—

"Vere magna et longe pulcherrima sunt etiam illa, profundissimâ sapientiâ hic exstructa opera tua, O JEHOVAH! quæ non nisi bene armatis nostris oculis patent! Qualia autem erunt denique illa, quæ, sublato hoc speculo, remotâ mortalitatis caligine, daturus es tuis Te vere sincero pectore colentibus! Eheu qualia!"

"Truly great and transcendently beautiful, O JEHOVAH! are these thy works even here below. Framed they are in profound wisdom, disclosing all their charms only to our lens-aided eyes! How grand, then, will those be which—when this glass has been removed in which we see darkly—when this mist of mortality has been scattered—Thou art pledged to reveal hereafter to thy servants, that have worshipped Thee here in sincerity and truth! Ah me! how grand!"

ment partly by their roots from the soil, and partly from the air by their leaves. There can be no doubt as to their being nourished by their roots, for manure applied to the roots soon tells upon the plants; and they extend far in search of nourishment. Many think, however, that sea-plants derive all their nourishment from the water, or the air when exposed to it, by means of their general surface, and that the roots are scarcely of any use except to attach them to the rock or other substance on which they grow. In confirmation of this opinion they mention that *Sargassum bacciferum*, or Gulf-weed, which floats in such mighty masses in some seas, not only seems to grow in this floating state, but that this floating piece of *Sargassum* has never been found attached by roots. It may nevertheless yet be found in a young state attached by roots, though it may still be true that it can grow in this floating state. Take a branch of *Cladophora glomerata* from a plant adhering by roots to a stone in fresh water, and place the branch in a vase of fresh water, and it will grow and increase considerably. Though a land-plant suspended in the air without getting nourishment by the roots would soon wither, yet you know that some of the *Epidendra*, air-plants as they are called, are exceptions from this, for if they are hung up in a room they will grow for years, though they get no nourishment except what they derive from the air by their surface. We doubt not, then, that in some instances, such as the floating *Sargassum,* sea-plants grow without deriving any nourishment by roots, but we are disposed to regard such cases as exceptions from the general rule. We cannot persuade ourselves that they draw no nourishment from the substances to which they are attached. They show a decided preference for certain kinds of rocks and Algæ, and their growth is more luxuriant when they are found on those they prefer. Limestone seems a great favourite with some, and the abundance of limestone rock on many parts of the Irish coast, is probably one reason why many of the Irish specimens put to shame the English and Scottish dwarfs of the same species. It is well known also that there are several species that grow both on rocks and on large Algæ, and it is as well known that, in the same habitat, those growing on rocks and those on Algæ differ considerably in appearance, and this seems to be owing to different nourishment by roots, as other circumstances are the same.

A.

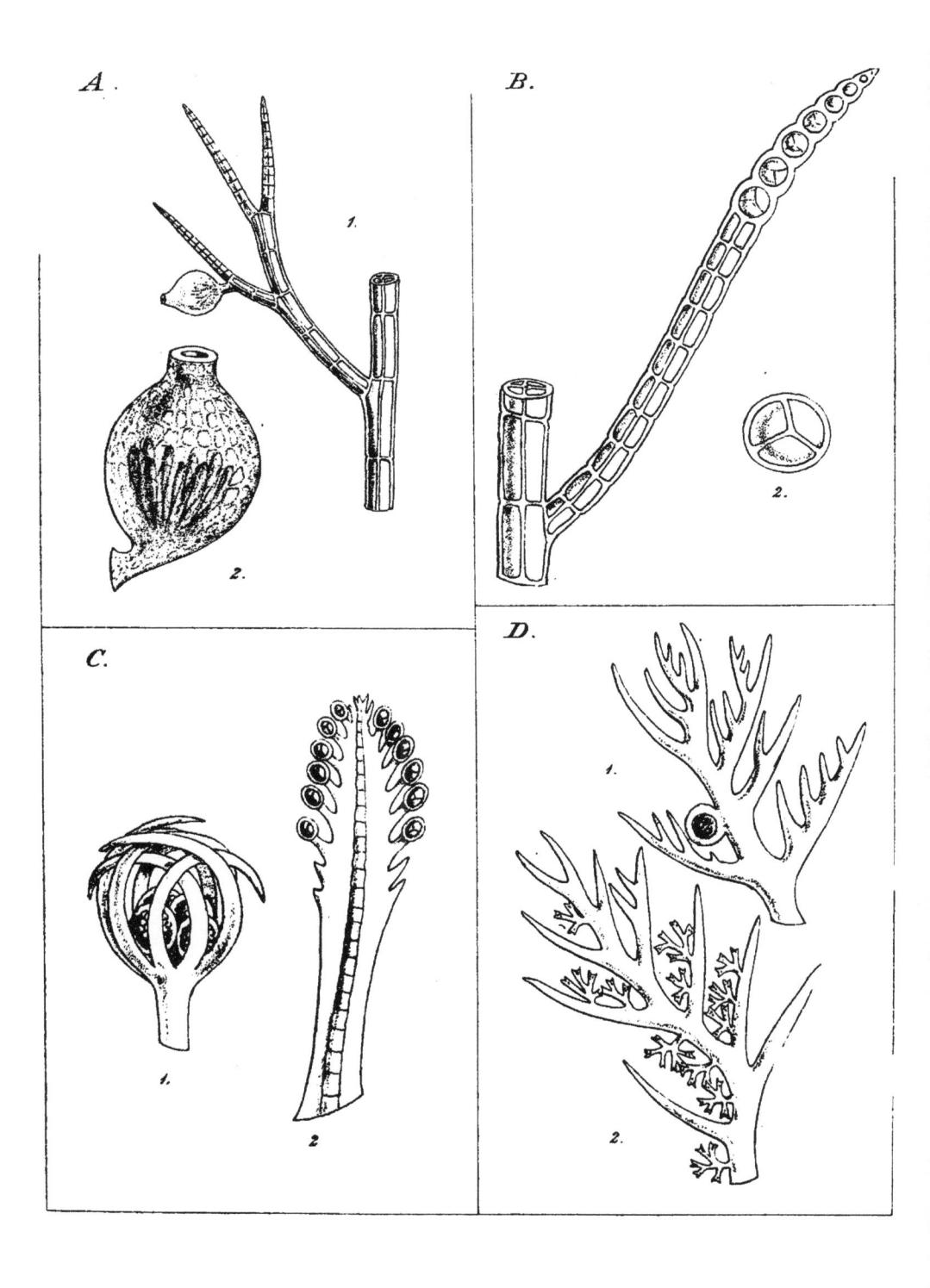

1.

2.

B.

1.

2.

C.

1.

2.

D.

1.

2.

CHAPTER V.

FRUCTIFICATION OF SEA-WEEDS.

"Dices * * * quod temporis et studii in eam impendendi jacturam compenset? Respondeo, Voluptas illa et delectatio innocens et honesta quæ ex operum Dei contemplatione oritur."—*Raius*.

> "Thy desire which tends to know
> The works of God, thereby to glorify
> The great Work-master, leads to no excess
> That reaches blame, but rather merits praise,
> The more it seems excess; * * *
> For wonderful, indeed, are all his works,
> Pleasant to know, and worthiest to be all
> Had in remembrance always with delight."—*Milton*.

WHEN Dr. Patrick Neill wrote his excellent article on *Fuci* about twenty years ago, he said that till within these few years the fructification of the *Fuci* was little known, and even yet it is not by any means understood. Since that period, however, very great progress has been made in this department. The importance of the fruit in determining the genus or species to which a plant belongs, is now fully acknowledged, and our most distinguished Algologists, aided by vastly improved microscopes, having prosecuted their researches with great diligence, have, both by accurate descriptions and by magnificent figures, given us much more accurate ideas respecting the different kinds of fructification. Many Sea-weeds are furnished with a double system of fructification, called for convenience *primary* and *secondary*, though it is acknowledged that the latter is as capable of producing a new plant as the former. The greater part, if not all, of the red series is furnished with this double fructification, the primary, or capsular, on one plant, and the secondary, or granular, on another plant. Though the capsules of two different genera and species bear no resemblance to each other, yet there are many modifications of shape, so that they are at once like and unlike, and to become acquainted with the minor differences requires time and attention.

That our young friends may at one glance have a view of several of the forms which the fructification of Algæ assumes, we have introduced some figures of them, chiefly taken from the elegant plates in Dr. Harvey's 'Phycologia Britannica' and Dr. Greville's 'Algæ Britannicæ,' as both these gentlemen kindly allow me to avail myself of their works. At letter *A*, fig. 1, there is a small portion of a branch of *Polysiphonia urceolata*, with a capsule considerably

E.

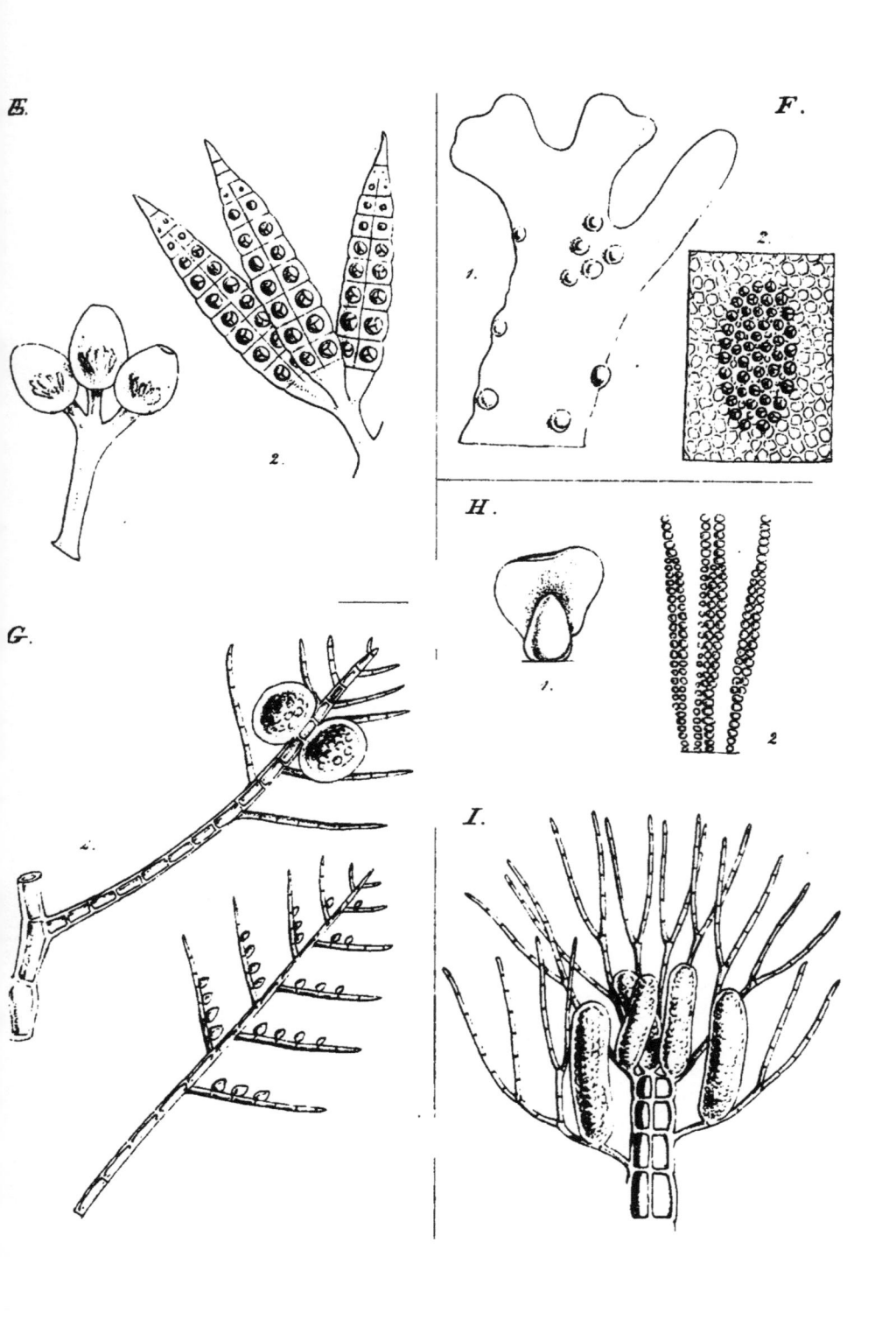

F.

1.
2.

H.

1.
2.

G.

I.

magnified; and fig. 2, a capsule of the same greatly magnified. This elegantly urn-shaped capsule is also called a *ceramidium,* a name applied to a capsule furnished with a terminal pore or opening, and containing a tuft of pear-shaped spores. Now by looking at the greatly magnified capsule or ceramidium, you will see that it is open at the top, and that it contains the tuft of spores or seeds. This *Polysiphonia* with its *capsular* fruit is very common everywhere.

We next exhibit, at *B,* the granular fruit of this same *P. urceolata,* no less common. Figure 1 is a portion of a branchlet, with the granules imbedded in the upper half; and these are called *tetraspores,* because, when much magnified as in fig. 2, they are found each to consist of four spores or seeds. You observe also in the branchlet, the joints and the longitudinal tubes or siphons.

At *C,* is the representation of two kinds of fructification found on a very common, but beautiful plant, *Ptilota plumosa.* Fig. 1 is an *involucre* containing favellæ; fig. 2, a *lacinia,* with tetraspores on short pedicels: magnified.

At *D,* two kinds of fructification of the beautiful *Plocamum coccineum* are represented. Fig. 1, a branchlet with a lateral tubercle containing spores; fig. 2, a branchlet with *stichidia* containing tetraspores : all magnified.

At *E*, are figures of two kinds of fructification observed in abundance on *Odonthalia dentata* in early winter. Fig. 1, capsular fruit; and fig. 2, *stichidia*.

On flat membranaceous Algæ, two kinds of fructification are common. At letter *F*, fig. 1 represents a portion of the frond of *Rhodomenia bifida*, with imbedded *tubercles*, called also *coccidia*. These contain spores or seeds : magnified. Fig. 2 represents a portion of the frond of *Nitophyllum punctatum*, or rather, it is one of the *sori* with which the frond is spotted. These *sori* consist of an assemblage of tetraspores : magnified.

The genus *Callithamnion* has two kinds of fructification. At the letter *G*, fig. 1 represents the *capsular* fruit of a *Callithamnion*, sessile on the upper side of the pinna. Fig. 2, a branchlet with a bilobed *favella*. The favellæ have sometimes only one lobe.

There is another kind of fructification, called by Agardh *nemathecia*. These *nemathecia*, or warts, are concealed under leafy processes; and at letter *H*, fig. 1, is a wart on *Phyllophora rubens*, with one of the leaves removed to show the nemathecia. Fig. 2 represents the moniliform filaments, much magnified, which compose the nemathecium.

The only other kind of fructification of which I shall give a figure, is that called *antheridia*. These are not uncommon. They are found in so great abundance on *Polysiphonia fastigiata* at times, as to give a yellow colour to the plant quite observable by the naked eye. At letter *I*, fig. 1 represents the tip of a filament of *Polysiphonia fibrata* with *antheridia* among the spiral fibres.

"The seed," says Professor Harvey in his Manual, "appears to consist of a single cellule or bag of membrane, filled with a very dense and dark-coloured granular or semifluid mass, called the endochrome. This seed, or germination, produces a perfect plant, resembling that from which it springs. Its growth may be watched from the commencement, when what we may call the ovule, or germ of the future seed, begins to swell. But nothing whatever has been ascertained that throws the smallest light on the process of fecundation."

Lindley, in his 'Vegetable Kingdom,' says: "The younger Agardh has expressed his deliberate opinion that in the Rose Tangles (his *Floridea*), organs analogous to sexes are present. 'I am very much inclined,' he says, 'to adopt the opinion that the two sorts of fructification observable among them are the first attempts at the agency which in higher plants perform the office of sexes, without, however, having their qualities established, and each capable of producing a new plant, without the aid of the other.'

"M. Decaisne seems also to have altered his opinion on this subject, for he and M. Thuret now describe what they suppose to be sexual organs in *Fucus serratus* and other species, to which they even apply the Linnean names Monœcions and Diœcious. They describe the conceptacles of the males as being filled with articulated filaments, bearing numerous antheridia in the form of vesicles containing red granules. 'These antheridia are expelled by the orifice of the conceptacles; if we examine them with a microscope, we see issue from one of their extremities transparent somewhat pear-shaped bodies, each enclosing a red granule. Every one of such bodies is furnished with very thin cilia, by means of which it moves with very great activity.' Such bodies are regarded as analogous to the spiral threads of mosses and other cryptogamic plants. Indeed, according to M. Thuret, such threads are also furnished with ciliary locomotive organs. But what proof is there," inquires Dr. Lindley, "that these curious bodies are pollen?"

There are remarkable circumstances connected with the fructification of many of the *Chlorospermeæ*, but I shall reserve what I have to say on that subject till I come to treat of some of the fresh-water Algæ.

CHAPTER VI.

GEOGRAPHICAL DISTRIBUTION.

" Ποντιων τε κυματων
ἀνηριθμον γελασμα."—*Æsch. Prom.*

"Countless laughing gleams of deep sea-weeds."

"When up some woodland dale we catch
The many-twinkling smile of Ocean,
Or with pleased ear bewildered watch
His chime of restless motion;
Still as the surging waves retire,
They seem to gasp with strong desire;
Such signs of love old Ocean gives,
We cannot choose but think he lives."
The Christian Year.

To discuss this subject properly, even though I were equal to it, would require far more space than this little work could allow. After a few remarks, I shall gladly refer to Dr. Neill, Dr. Greville, and Dr. Harvey, and to the able essay on the subject by Lamouroux, to whom they express their obligations.

Algæ, or Sea-weeds, have a wide geographical range, for

I suppose wherever there is sea, sea-weeds of some kind are found. To a considerable extent, they seem to obey the same laws as land-plants. Every zone presents a peculiar system of vegetation. It is said that after a space of twenty-four degrees of latitude, a nearly total change is produced in the species of organized beings; this, it is thought, is very much owing to temperature. It is very interesting to observe the different appearances that Sea-weeds exhibit in different seas, and also in different depths of the same sea. How very dissimilar are Algæ of the Arctic Seas from those of Australia and New Zealand! How remarkable are the immense masses of floating Sea-weeds, such as those proverbial wanderers the *Sargassa*, called also Gulf-weed!

Great accumulations of Sea-weed are found floating on each side of the equator. The *Mar do Sargasso* of the Portuguese stretches between the 18th and 32nd parallels of north latitude, and the 25th and 40th meridians of west longitude. It is often called the Grassy Sea, for in it the surface of the ocean, for several days' sailing, is literally covered with plants. Barrow mentions them in his 'Voyage to Cochin China'; Humboldt also mentions them in his 'Personal Narrative.' The most extensive bank is near the Azores: vessels returning to Europe from Monte

Video or the Cape of Good Hope cross it. Columbus, it is well known, encountered most extensive banks, that impeded in some degree the progress of his vessels. If, however, we consider the general distribution of the three great series, namely, the olive, the red, and the green, we learn that the first increases as we approach the tropics; that the second chiefly abounds in the temperate zone; while the green Algæ greatly predominate in the Polar seas. The Creator has assigned to each the peculiar kinds best suited to the climate and other circumstances.

In the Botany of the Antarctic Expedition of Sir James Clarke Ross, Dr. Hooker has given a miniature representation of a submarine forest of *Lessonia* and *Macrocystis*, inhabiting the south circumpolar seas, in which the trunks of the former, growing under water in an erect position, are from five to ten feet in height and of the thickness of the human thigh. Thousands of these aquatic trees, uprooted by the currents, are not unfrequently mistaken for drift-wood, and collected for fuel. As an example of the gigantic growth of Sea-weeds in this region, Dr. Hooker observes that the *Macrocystis luxurians*, in its horizontal growth at the surface of the ocean, ranges between 200 and 700 feet in length, and that at the Falkland Islands the beach is lined for miles with entangled cables of this plant, much thicker than the human body.

We have much to learn respecting the inhabitants of the deep water, animal as well as vegetable, especially in foreign climes. The surpassing richness and variety of submarine scenery in favoured seas, must have an almost overpowering effect on the intelligent observer, during a voyage of scientific research. Respecting submarine scenery in general at *Macao* in China, Mr. Adams, of the Samarang, writes:—

"Dendritic zoophytes, having their branches loaded with coloured Polypi, like trees covered with delicate blossoms, richly uprose from the clear bottom of the bay, distinct and characteristic in their specific forms, and contrasting strangely and powerfully with those most apathetic and stone-like combinations of the plant, the animal, and the rock,—the Madrepores, the Millepores, and the Nullipores. Flat, and immoveably extended on the sand, in the bare spots between the Corallines, were impassive, large, blue, five-fingered *Asterias*; and, crawling with an awkward shuffling movement, like an *Octopus*, were numbers of the slender *Ophiuræ*, with their snaky arms, groping their way among the weeds, and striving to insinuate their writhing forms among the coral masses. Fixed flower-like *Actiniæ* were expanding their fleshy petals on the rocks; the slender *Nereis*, the long-armed *Comatula*, and the languid, slow-moving *Holothuria*, together with numerous fish and crustaceans, contributed to prove that Nature is ever weaving the subtile woof of existence beneath the surface of the waves."

CHAPTER VII.

RANGE OF BRITISH SEA-WEEDS.

"O Lord, how manifold are thy works! In wisdom hast thou made them all. The earth is full of thy riches: so is this great and wide sea."— *Psalm CIV.*

"Almighty Being!
Cause and support of all things! Can I view
These objects of my wonder,—can I feel
These fine sensations, and not think of Thee?
Thou who dost through the eternal round of time,
Dost through the immensity of space exist
Alone, shalt Thou alone excluded be
From this thy universe?"—*Stillingfleet.*

Soil and climate are the chief causes that influence the distribution of land-plants; and we doubt not they are also the chief causes that operate in the distribution of sea-plants. Their influence, however, is less in sea-plants, because they receive much less nourishment by means of their roots, and also because the temperature of the sea is much less variable than the temperature of the air. Some of our British Sea-weeds are so widely spread in various climates and seas, that they may be regarded as citizens of the world, and therefore are little likely to be influenced by any perceptible difference of temperature betwixt the most southerly and the most northerly portions of our British seas. As we do not expect to hear the nightingale in the northern forests of Scotland, or to see the Alpine plants of Ben Nevis blooming in the rich sheltered plains of the south of England, so there are Sea-weeds that are abundant in England that are never found in Scotland, and others that are plentiful in our Scottish seas that are never met with on the genial shores of Devonshire. Dr. Greville, in his admirable 'Algæ Britannicæ,' observes:—

" On the shores of the British Islands it is easy to perceive that some species,—*Gelidium corneum, Phyllophora rubens,* and *Sphærococcus coronopifolius,* for example,—become plentiful and more luxuriant as we travel from north to south; and, on the other hand, that *Ptilota plumosa, Rhodomela lycopodioides,* and several others, occur more frequently, and in a finer state, as we approach the north. *Odonthalia dentata* and *Rhodymenia cristata* are confined to the northern parts of Great Britain, while the *Cystoseiræ, Fucus tuberculatus, Haliseris polypodioides, Rhodomenia jubata, R. Teedii, Microclodia glandulosa, Chodymenia pinastroides, Laurencia tenuissima, Iridæa reniformis,* and many others, are confined to the southern parts. Others, again, such as the *Fuci* in general, the *Laminarieæ*, many *Delesseriæ*, some *Nitophylla, Laurenciæ, Gastrideæ,* and *Chondri*, possess too extended a range to be influenced by any change of temperature between the northern boundary of Scotland and England."

Even in those that are found only in the north or in the south, we are not always able to account for their abundance in some seasons, and for their scarcity in the same localities in others. The case is the same with Mollusca and Zoophytes. Nor can we always account for the abundance of the last-mentioned very interesting creatures in some localities, and their utter absence in others where the climate is quite the same.

Dr. Neill mentions that on our shores Algæ generally occupy zones in the following order, beginning from deep water:—*F. filum, esculentus,* and *bulbosus; F. digitatus, saccharinus,* and *loreus; F. serratus* and *crispus; F. nodosus* and *vesiculosus; F. canaliculatus;* and, last of all, *F. pygmæus,* which is satisfied if it be within reach of the spray.

CHAPTER VIII.

THE USES OF SEA-WEEDS.

" How wondrous is the scene ! where all is formed
With number, weight, and measure ! *all designed
For some great end* ! Where not alone the plant
Of stately growth, the herb of glorious hue,
Or foodful substance ; nor the labouring steed,
The herd and flock that feed us ; nor the mine
That yields us stores for elegance and use ;
The sea that loads our table, and conveys
The wanderer, man, from clime to clime, with all
Those rolling spheres, that from on high shed down
Their kindly influence ; nor these alone
Which strike e'en eyes incurious, but each moss,
Each shell, each crawling insect, holds a rank
Important in the scale of HIM who framed
This scale of beings ; holds a rank, which, lost,
Would break the chain, and leave a gap behind
Which Nature's self would rue ! "—*Stillingfleet.*

No person, now-a-days, will venture to pronounce Sea-weeds useless. Had we known no economical purposes to which they could be applied, still we should have been bound to adore the goodness of God in clothing the rocks and channel of the sea with so much beauty. Much of this beauty may be seen from dry land, but submarine scenery is viewed to far greater advantage during the leisurely movements of a row-boat in a summer day. How rich and various their colours! and how gracefully do the branches of these submerged forests wave in the deep! And though there are no birds to enliven the scene, there are thousands of fishes and crustaceans, whose movements are not less interesting than those of birds. If few see these things, does it not magnify the condescending kindness of God that he makes so rich a provision for the happiness of these few?

> "Full many a gem of purest ray serene
> The dark unfathomed caves of ocean bear;
> Full many a flower is born to blush unseen,
> And waste its sweetness in the desert air."

And yet even in the desert air the sweetness is not wasted, for the "little busy bee" can enjoy it; nor is its beauty always unseen by man. Who has not heard of the effect which the sight of a little plant produced on Mungo Park, the African traveller? When, ready to perish with hunger and fatigue, he had laid himself down to die in the desert, even in these circumstances he could not help contemplating a little moss (*Dicranum bryoides*) which attracted his attention, covering the ground on which he lay; and so much struck was he with its exquisite beauty, that he felt constrained to say, " Can He who gave so much beauty to that diminutive plant, and who thus cares for it in the desert, be forgetful of me?" Cheered by the thought, he started up, pushed on with renewed vigour, and soon reached the habitations of men.*

We may remark, however, that if Mungo Park had not been an acute observer, the beauty of this little plant would have been unseen by him, even when it was close to his eyes. The eye requires training. Many have eyes and yet no eyes, because they have not been accustomed to make a right use of them. This little plant, which had so cheering an effect on the observant traveller, is not uncommon in our own country; and a bank covered with it in a state of fructification is one of the most lovely spectacles on which we can look. Though small, it is not so minute but that its beauty can be perfectly seen by the naked eye; and yet so little do we in general attend to those things that have no great magnitude to recommend them, that were we to

* Sir William J. Hooker, in his ' British Flora,' says : " The moss which engaged Mungo Park's attention so much in Africa as to revive his drooping spirits when sinking under fatigue, was this species (*Dicranum bryoides*); as I have ascertained by means of original specimens given to me by his brother-in-law, Mr. Dickson."

take at random ten thousand persons from either town or country, and march them along a bank carpeted with myriads of this beautiful fork-moss, when every frond is surmounted by its lovely fructification, not ten out of the ten thousand would take notice or say, "Is not that charming?" The majority would also see it, and *not* see it; yet if their attention were specially directed to it, they would almost all wonder that they had not before greatly admired it.* We deprive ourselves of much innocent pleasure if we are not observant of the wonderful works of God. And how much may the mind be enlarged by the devout and intelligent contemplation of what He has so wonderfully made! "The more we extend our researches," says Dr. Greville, "into the vegetable kingdom, the more will every susceptible mind be excited to proceed. We shall find the most delicate and elaborate processes in ceaseless progression on the mountains and in the valleys, the meadows and the recesses of our woods, on the rocks of our friths and seas—all subject to immutable laws. We shall find colours unrivalled, odours inimitable, and forms

* I may mention to my young friends that what is called *fructification* in mosses and sea-weeds and other cryptogamic plants, answers the same purpose as flowers in phænogamous plants.

exhaustless in variety and grace, daily developed in the grand laboratory of nature, demanding only to be seen to extort our unqualified admiration, and leading us irresistibly to contemplate the glory of that almighty Being from whom so many wonders emanate."

The direct and indirect uses of Sea-weeds already known are so numerous that we cannot venture to mention them all. There is a respectable class of wholesale dealers in Sea-weeds who are very far from despising them,—we mean our agriculturists. It is interesting, even on our Ayrshire coast, to contemplate the busy scene which the sea-shore exhibits after a breeze, when young and old are actively engaged in gathering up the treasure which the bountiful ocean has spread out for them. There we see the farmer's cart, and the cottager's loaded barrow; the industrious boy has also his bagfull, which he offers for a penny; and even children, though they have neither bag, basket, nor barrow, contrive to fasten a load on the broad-fronded Tangle, and using the stem as the pole, the little urchins drag it along in triumph. In the Island of Arran, opposite to the Ayrshire coast, the value of Sea-weed as manure is well known; and, lest there should be any dispute respecting its appropriation, the factor for the Duke of Hamilton very judiciously assigns to each who has an interest in the adjoining land, the portion of the shore from which at will he may collect Sea-weed. Great, indeed, is the avidity with which sea-wrack is gathered as manure on all parts of the British coast. In Ireland it is still more valued, as it is the chief manure for thousands of acres of potato-ground. "On many of our coasts," says Dr. Harvey, in his Manual, "as along the west coast of Ireland, the poorer classes are almost entirely dependent for the cultivation of their potatoes on the manure afforded by their rocky shores and frequent gales." And Mr. W. Thompson, of Belfast, in the 'Annals of Natural History,' says: "Of this we had a notable example at the town of Galway, some years ago. Turf-boats were discharging their cargoes of Sea-weed at the quay, and on inquiry whence it was brought, we learned that it was from Slyne head, a place distant between fifty and sixty miles, and that some of the purchasers were, for the purpose of manure, about to convey it inland thirty miles!"

Nor is it in Britain alone that the value of Sea-weed is known. Inglis, in his 'Channel Islands,' gives a most interesting account of the-use made of this production of the sea in Jersey and Guernsey. The name which the inhabitants of these islands give to Sea-weed is *vraic*, evidently a corruption of *varec*, the French word for Sea-weed.

According to Inglis, collecting *vraic* is so important a matter in the Channel Islands, that the seasons when the operations of cutting and collecting it begin, are appointed by law. The seasons are two, usually the 10th of March and the 20th of July, and they continue about ten days. When the *vraicking* season has come, if a family be not sufficiently numerous for the work, they are joined by some of their neighbours, and the parties, consisting of eight, ten, or twelve, sally forth betimes from all parts of the island to their laborious, but cheerful work. Though a time of labour, it is also a season of merriment,—the "vraicking cakes," made of flour, milk, and sugar, are plentifully partaken of. On the cart which accompanies the party to the sea-beach, there is generally slung a little cask of something to drink, with a suitable supply of eatables. Every individual is provided with a small scythe to cut the weeds from the rocks, and with strong leg and foot-gear. The carts proceed as far as the tide will allow, and boats carry the *vraickers* to those more distant rocks which are not approachable in any other way.

"It is truly a busy and curious scene," says Inglis;

"during this season, at half-tide or low water, multitudes of carts and horses, boats and *vraickers*, cover the beach, the rocks, and the water; and so anxious are the people to make the most of their limited time, that I have often seen horses swimming and carts floating, so unwilling are *vraickers* to be driven from their spoil by the inexorable tide." The *vraick* is used as manure either fresh from the rocks, or after it has been burnt as fuel. Inglis mentions the remarkable fact, of which I was before ignorant, that in these islands *vraick* is the chief and almost the only article used as fuel. For this purpose it is collected at other times than the regular *vraicking* seasons, and then it consists of what has been detached from the rocks by the waves, and carried to the shore by the tide. At all times men, women, and children, but chiefly the latter, may be seen at this employment. They use a rake or three-pronged pitch-fork, and a wheel-barrow in which it is carried above high-water mark, to be spread out and dried. It makes a hot, if not a cheerful fire. Scarcely any other fuel is used in these islands: a little wood, though rarely, is mixed with it; and it is only on feast-days and family festivals that a coal fire is lighted in the best parlour.

Early in the morning, if a person is strolling abroad, he is apt to suppose that the Jersey farmer and his household have been astir before day-break, for the smoke is seen rising from the cottages. But the fires have been burning all night;—there would be no thrift in extinguishing the *vraick*-fires, as the consumption of fuel is the manufacture of manure, the burnt *vraic* answering better for their fields under crop, than the fresh *vraic*, which is employed as a top-dressing for their green-fields.

We see that Sea-weeds are very valuable as manure, and as fuel; are they not also employed as food? Though not so much used in this way as they once were, there are some kinds that are still much eaten by the Highlanders and by the Irish, and to some extent by the Scotch Lowlanders. "Wha'll buy dulse and tang?" was one of the euphonious cries which tickled my ears, when, from an inland part of the country, I came as a young student to the University of Edinburgh. It was much eaten by the Highlanders till it was supplanted by that nauseous herb, tobacco; and well would it have been for both purse and person, if they had continued to prefer it to a costly narcotic. Its very wholesomeness was one of its great recommendations; long before it was known to contain iodine, and before iodine itself was discovered, it was thought very efficacious as a sweetener of the blood, and in warding off, or curing, scorbutic and glandular affections. The stick which is chewed by the inhabitants of the Alps affected with *goître*, a disease of the glands of the neck, is said to be the stem of a kind of Tangle. There is, said Dr. Neill, a common saying in Stronza that " he who eats of the *Dulse* of Guerdie and drinks of the wells of Kildingie, will escape all maladies except black death." Even *pale* death *(pallida mors)* laughs at the prescriptions of the most skilful physicians, and *black* death, it would appear, is not in the least more merciful. A writer in the 'Quarterly Review' says, " *Dulse* to the Icelander is a plant of considerable importance. They prepare it by washing it well in fresh water and exposing it to dry, when it gives out a white powdery substance, which is sweet and palatable, and covers the whole plant. They then pack it in casks and keep it from the air, and, thus preserved, it is ready to be eaten either in this state, with fish and butter; or, according to the practice of wealthier tables, boiled in milk, and mixed with a little flour of rye." We doubt not that the white powdery substance which it gives out is *Mannite,* which we shall afterwards show is pretty abundant in several of our Sea-weeds. Cattle also are very fond of this Sea-weed, and sheep are said to seek it with such avidity that they are often lost, by going too far from land at low-water in search of it, and becoming surrounded by the returning tide. The Norwegians, therefore, call it *sou-soell,* and Bishop Gunner, translating this into Latin, calls it *Fucus ovinus,* Sheep's-weed.

But Dulse is not the only Sea-weed useful as food for cattle. In the Western Hebrides, *Fucus vesiculosus,* or *Lady Wrack* as it is often called, forms a considerable part of the winter food of cattle and sheep. Even in the island of Cumbrae, only a few miles from the coast of Ayrshire, the minister of the island told me that during winter his man-servant went regularly to the shore at every ebb-tide to cut from the rocks a fresh supply of this Sea-weed for his cows, who neither snuffed nor turned up their noses at it, but relished it and throve upon it.

But while the Icelanders make a savoury dish of Dulse, the Irish peasantry have taught us to make a truly delicious dish of another very common Sea-weed, well known by the name of *Carrageen* or Irish moss. This is *Chondrus crispus,* or, what answers the purpose equally well, *Chondrus mamillosus,* to which the name of *Gracilaria mamillosa* has now been given. These two plants abound on all our rocky shores; but it is called Irish moss, because it was first turned to account in Ireland. Being recommended as a palatable food, and particularly as light and nourishing for invalids, it became a fashionable dish; and the dried material sold at one time as high as 2s. 6d. per pound. It is now to be got at a much cheaper rate in apothecaries' shops, but as many would prefer a repast direct from the sea, we may mention that it is bleached in the same manner as linen or cotton, and when dry it can be kept for years. When used, a tea-cup full of it is boiled in water; this water, being strained, is boiled with milk and sugar and some seasoning, such as nutmeg, cinnamon, or essence of lemon. It is then put into a shape, in which it consolidates like blancmange, and when eaten with cream it is so good that many a sweet-lipped little boy and girl would almost wish to be on the invalid-list to get a share of it. There is a Chinese Sea-weed lately imported for commercial purposes, which is used very much in the same way as the Irish moss, and forms even a daintier dish: the native name is Agar-Agar, or Agal-Agal. It is thought to form a component part of the celebrated nests of a species of swallow, *Hirundo esculenta*,*

* The following is from that rich and interesting publication, 'The Voyage of H.M.S. Samarang:'—

"About the rocky parts of the coast of Borneo, the *Hirundo esculenta* swims backwards and forwards all day long, uttering its little cheerful chirp regarded as such a luxury that they sell for their weight in gold.

There were several kinds of Sea-weeds formerly used for food, that are not now much eaten, though they may still retain the specific name of *edulis*. Tastes change: what was eaten with relish by our forefathers may not always be regarded as a *bonne bouche* by their posterity. The time has been when the tongue of a porpoise was reserved as a special dainty for the Royal table; we suspect that the tongue of a stag, or peradventure of an ox, would be more to the taste of our good Queen Victoria. Some of our Sea-weeds that were once welcome at the festive board, we surmise get no higher than the tribes whose names they rejoice to bear, such as *Swine-tang*: nor are they without their value, if they can furnish a feast, even to a greedy porker. Some, however, are still prized at the tables of the great. *Porphyra* is gathered for culinary purposes, in

as it eagerly pursues its insect prey. I have taken the nests in nearly every state from the sides of hollow caves, where they adhere in numbers to the walls like so many watch-pockets.

"The Malays frequently assert that the nests are formed from the bodies of certain sea-snakes, but there is no doubt that Agal-Agal, a marine cellular plant, is the material employed in forming those much-prized eatable nests."

England under the name of *Laver*, in Ireland under the name of *Sloke*, and in Scotland as *Slaak*. In Scotland it is seldom used, except as a luxury by the affluent under the English name of *Laver*. It is prepared in different ways: sometimes it is boiled for hours, and, when reduced to a pulp, eaten with lemon-juice; at other times it is well boiled and seasoned with spices and butter. So far as our own experience goes, it requires them all; with these appliances, however, it is tolerable, verifying the good Scottish proverb, " If you boil *stanes* in butter, you may sup the *broo*."

Some, however, are of opinion that by proper management our Sea-weeds might yield in time of need even the necessaries of life. A distinguished and well-known chemist of our own country says, in the 'Edinburgh Philosophical Journal,' that a gum might easily be procured from them, that would serve all the purposes of gum arabic, and which, by reason of its cheapness, might be applied to a host of other purposes. How is it, argues he, that gum is so little used as an article of diet in this country, seeing that its nutritious qualities are so well attested by the fact, that the Moor of the desert can subsist on six ounces of gum a day for weeks together? Why is it that so many of our countrymen bear the signs of famine in their eyes, and are constantly exposed to so many moral and political evils, while treasures of such wholesome food thus lie scattered in the greatest profusion on our shores?

In a commercial point of view, our British Sea-weeds rise to national importance on account of *kelp*, which is made from them, and which was much employed for many years in the manufacture of soap and glass, though of late chiefly valued on account of the iodine it yields. In the last part of the Proceedings of the Glasgow Philosophical Society, there is an excellent article by Mr. Glassford on the manufacture of kelp, from which my space will allow me to glean much less than I could wish. The rise in price of kelp about the beginning of the present century, from 3*l.* to 20*l.* or 22*l.* per ton, caused the Highland proprietors to devote much of their attention to the manufacture of kelp; and not only carefully to collect the Sea-weeds that the rocks yielded, but to make artificial plantations in some cases for Sea-weeds, by covering the beach visited by the tide, with large stones, which soon bore a crop convertible into kelp. The supply could scarcely keep pace with the demand; the manufacture was pushed to the farthest limit, and for years continued to flourish. Barilla, however, from abroad entered the market, and reduced the price of kelp to about ten guineas a ton. Barilla contained more soda, and was preferred even at a higher price. The duty, however, was taken off barilla and salt, and then kelp fell to less than 3*l*. per ton. It then yielded nothing directly to the proprietor; but the manufacture was to a certain extent continued; for it afforded employment to the tenants and helped them to pay their rents, and kept them from being a burden.

Highland estates, that had become so valuable during the flourishing state of the kelp manufacture, now experienced a wonderful depression. Mr. Wilson, in his very interesting and amusing account of his Voyage round the coast of Scotland and the Isles in 1842, says, that in 1812, in the island of North-Uist, the clear proceeds from kelp alone, after deducting all expenses, was 14,000*l*. and fell little short of that sum for several years after; but that the alteration of the law regarding the duty on barilla, reduced the income of that island and its dependencies from 17,000*l*. to 3,500*l*. When M'Culloch visited the Hebrides in 1818, the total product of kelp from these islands was estimated at 6,000 tons, which at 20*l*. a ton must have realized the sum of 120,000*l*. At present there is reason to believe that not much over 3,000 tons are annually manufactured, from which not above 6,000*l.* are realized, after deducting the wages of the kelpers and the expense of necessary apparatus. This small remuneration, however, is owing in part to the Highlanders' perseverance in manufacturing their kelp from the *yellow wrack* (*Fucus nodosus*), which, from growing in shallow water, and being less thoroughly a marine plant, yields much less iodine than the kelp manufactured from the *black wrack,* such as *Laminaria digitata,* the Great-stemmed Tangle ; and *Fucus serratus,* the Serrated Sea-weed. Irish kelp prepared from *drift-weed* from deep water, is rich in iodine, yielding above twelve pounds per ton, and, consequently, carefully and honestly prepared Irish kelp brought in some cases 10*l.* per ton, in 1845, when Highland kelp would not bring above the half of that sum. The rise which, after twenty years of depression, took place lately in the price of kelp, was owing to a great additional demand for iodine.

Who knows what virtues we may yet discover in the drapery of the deep? The book of nature is like the book of grace, the wonders they contain must be "sought out." Iodine, it would appear, contributes in some way to the health of marine plants, for they all have the power of extracting it from the waters of the deep. Or is this power given them for the good of living creatures, and especially of man, who often derived benefit from it in the use of Sea-weeds, though he knew not of its existence? And has its existence and the way of extracting it been discovered so late in the day as 1812, to make us grateful for blessings unconsciously received, and to stir us up to more diligent research into God's works of nature, by the rich remuneration so unexpectedly bestowed? It will give us some idea of the value of the kelp and iodine manufactures, when we state that from July 1845 to July 1846, it is calculated that upwards of 10,000 tons of kelp were manufactured on our British shores, which, on an average of 5*l.* per ton, would amount to 50,000*l.*

It would be far from uninteresting to tell how the weed is collected by hardy fellows, leading a kind of amphibious life, every day drenched in sea-water, and not unfrequently deluged with rain, constantly occupied for three months in collecting and drying the weeds, and reducing them to kelp in the kiln; their almost only food during all that time being their hastily-prepared meals of oatmeal porridge with buttermilk, or treacle and water, followed by a bannock and a draught of water from the crystal brook; yet on that simple diet they continue healthy and hearty.

> "Sons of the rock and nurselings of the surge,
> Around the kiln, their daily labours urge,
> O'er the dried weed the smoky volume coils,
> And deep beneath the precious *kali* boils."*

But we pass over all this, and follow the *material* to Glasgow; and it will give some idea of the employment that it there gives, when we state that there are at present twenty establishments in Glasgow, some of them very extensive, for the *lixiviation* of kelp and manufacture of iodine, &c., working up in Glasgow alone about 3,000 tons a year.

From the ample materials furnished by Mr. Glassford, who has had opportunities of obtaining great practical as well as scientific knowledge of all the processes from first to last, I shall select some brief notices. The object of the chemical manufacturer, or *lixiviator*, as he is called, is to separate the various salts which the kelp contains. The most insoluble are those which are first separated, consisting of the *Sulphate* of *Potash*, the *Carbonate*, *Muriate*, and *Sulphate* of *Soda*, and the *Muriate* of *Potash*. The most soluble remain in the solution. In the solution, the *Iodides* and other very soluble salts are found, and it is from this liquor, called the *mother liquor*, that *Iodine* is extracted. This is easily done

* Is it necessary to tell our Scottish readers, that *kali* does not mean *kale*?

by the employment of manganese and sulphuric acid, which react on each other and produce oxygen, which again combines with hydrogen, and liberates the iodine, till now in union with the hydrogen. The iodine escapes from the liquor as a most beautiful violet-coloured vapour, which on cooling condenses into a black, solid, crystalline body, in large glass balloons arranged for the purpose.

Iodine is extensively used as a medicine, combined with potassium, with mercury, arsenic, &c. It is used also as a tincture dissolved in alchol, and in combination with oils it is much used as a liniment. Till some way of fixing it is discovered, it is too evanescent to be used in dying or calico-printing. In one respect the vapour of iodine is more wonderful than the mysterious mist that we have all read of, which, when liberated from the sealed box by the fisherman, consolidated and became a gigantic genie. Our violet mist, which, like that of Arabia, has been dragged from the sea, when artfully employed by man, can make the sun a portrait-painter. The sun is a painter from the beginning; but the ingenuity of man, availing itself of iodine as an obedient yet powerful genie, can make the sun a limner. My young friends know that for the wonders of Calotype and Daguerreotype, the vapours of iodine are indispensably necessary.

I dare not enlarge by telling how much use is made of the materials derived from kelp in the manufacture of soap, of alum, of green bottle-glass, &c., I shall merely mention that even the *kelp waste* is serviceable as manure. And if the *waste* be serviceable—how much more the kelp itself, if used in a pulverized state! They who know its component parts are confident that it would be more nutritious as the food of plants, than many manures that are purchased at a high price and brought from afar.

Before closing a statement of the direct uses of Sea-weeds, we may quote the words of Professor Burnett in his 'Outlines of Botany.' "*Alga inutilis,* exclaims an ancient poet; *vilior algâ est,* in a tone of contumely he adds; *refunditur alga,* repeats another Bard. The sea itself spurns forth the nathless flag—that flag the gathering of which for years enriched both peer and peasant on our northern coasts;—the very flag that now affords the iodine, which really does relieve that *evil,* which the *manus regalis,* the boasted royal touch (if it ever benefited the superstitious) so long has failed to cure."

Time and space would fail us were we to attempt to enumerate all the uses of Sea-weeds, but we shall not enlarge much more. I cannot, however, forbear quoting the following passage from the very interesting Voyage of the Samarang lately published. In the island of Borneo, "Tanjong Agal-Agal derives its name from the Sea-weed of that name, which is collected in large quantities upon these reefs, extending nearly two miles towards Batommande. There are several species of this *Fucus*, all soluble in water, forming a very nutritive mucilage, which, when mixed with acid, fruit, or made into jellies, produces a very grateful beverage for invalids. It forms a considerable article of trade with the Chinese, particularly in the northern provinces of Chin-chew, where it is manufactured into a bright substantial transparent yellow jelly, and is sent in boxes of about ten pounds each to Canton. The gum or paste made from it, is supposed to possess the advantage of being unpalatable to insects and worms. It is from this gum that their fancy lanterns are fabricated, by spreading it over gauze skeletons; it thus resembles, and is very frequently mistaken for, highly transparent horn. It is peculiarly brittle, even more so than glass, cracking under very slight changes of temperature."

We doubt not that it is this Agal-Agal of which a kind friend has sent us specimens from Glasgow under the name of Agar-Agar. The specimens bore a great resemblance to our British Alga, *Gracilaria compressa,* and was probably *Gracilaria lichenoides,* though several species and perhaps genera answer the same purpose, as is the case with our native mucilaginous Algæ. Some say that *Gigartina tenax* is the kind used by the Chinese for glue and fancy lanterns. I learn from Glasgow that Agar-Agar began to be imported about two years ago, under the belief that it would prove highly serviceable in many respects. The main object in view, however, was to employ it as a substitute for sago flour, which is used by manufacturers to starch and stiffen webs, and for this, on account of its gelatinous qualities, it seemed admirably fitted. It has not yet been sufficiently tried. The difficulty of completely dissolving it, has hitherto been the obstacle to its extensive use. Were this overcome, it would be of great service. As it is not costly, it may come into repute as an agreeable article of food. We have already mentioned that the blanc-mange which is formed from it, is exceedingly palatable, and, being the same in substance, must have all the nutritious virtues of the celebrated Swallow-Nests so much prized by the Chinese.

Even in our own most common Sea-wrack there are substances which may yet be turned to good account. One of these is *Mannite,* the characteristic principle of *Manna,* which my friend Dr. John Stenhouse has detected in many of our coarse Sea-weeds, but in greatest abundance in *Laminaria saccharina,* which we doubt not took its specific name from this circumstance. A quantity of this Sea-weed was by Dr. Stenhouse repeatedly digested with hot water, which formed with it a brownish, sweetish, mucilaginous solution. When evaporated to dryness on the water-bath, it left a considerable quantity of a saline semi-crystalline mass; this was reduced to powder and treated with boiling alcohol, by which a considerable portion of it was dissolved. The alcholic solution, on cooling, became nearly solid from the quantity of long transparent prismatic crystals with which it was filled. When purified by a second crystallization these were deposited in large hard prisms of a fine silky lustre. By analysis it was found that this was *Mannite*. The quantity of Mannite contained is very considerable: one thousand grains of the Sea-weed treated in the way described gave about 12 per cent. of Mannite. It is very beautiful—as purely white as loaf-sugar, and almost as sweet. Since I wrote the above I have examined and tasted Mannite which I got from Dr. Stenhouse about four years ago, and it is as white and sweet as ever. Surely some use may be made of this sweet marine treasure. No doubt it has been serving some good purpose already, were it only by sweetening the repast and adding to the happiness of the multitude of God's creatures that live on Sea-weeds in the deep.

HE who by a word clothed the rocks and channel of the sea with so much riches and beauty, and who said respecting the miraculous supply of food in the wilderness, " Gather up the fragments that remain, that nothing be lost," allows not even the fragments of Sea-weed to be lost. Man has learned the wisdom of gathering up those that are within his reach, but a kind Providence allows not to be lost the immense masses that are buried at the bottom of the sea. HE who in primæval ages stored up the remains of ancient forests, converting them into coal for the succeeding generations of the children of men, to give light and heat in their habitations; at the same time stored up the wreck of marine matter in the form of stone of which the palace and the cottage might be built. We cannot examine a limestone quarry without seeing that it must have been consolidated in the depths of the sea. Our marble jambs are a mixed mass of marine organic remains. Though Sea-weeds, being more perishable than the shelly coats of animals, are less frequently observed in marble or limestone, yet they are occasionally seen, and about a year ago in a limestone quarry near Ardrossan I saw numerous dark impressions of a large Sea-weed, resembling *Halidrys*. Nay, we have practical proof that the *disjecta membra* of Sea-weeds, buried in the mud, are well fitted to contribute to its consolidation. Mrs. Marshall, a talented and scientific lady of my acquaintance, after attending a meeting of the British Association, was led to ponder on the formation of rocks, and being a chemist, she thought she would try to construct them artificially. Many rocks, she observed, had been formed by stratification,—some at the bottom of fresh-water lakes,—others in estuaries, or at the bottom of the sea. What then must have been the component parts of the latter—of limestone, for instance? They are full of the organic remains of Mollusca and Zoophytes, whose habitations were formed of carbonate of lime, which they had the power of extracting from the sea. These shelly abodes of Mollusks, with the polypidoms of Zoophytes, such as corals, would often be broken in pieces and ground into powder by the mighty turmoil of the sea, and subsiding from time to time, would form strata of calcareous mud in which shell-fish and other marine animals would be buried. Along with these, there would constantly be mixed immense masses of Sea-weeds torn by storms and currents from the submerged rocks.

When these had hardened in process of time, and had become solid stratified rock, by great convulsions they were upheaved and kindly brought within the reach of man. Since it is evident, thought she, that both animal and vegetable matter must be copiously mixed in the soft mass converted into rock, may not such substances be in a manner necessary in such formations? And if I can hit on the right proportions, may I not, in my little laboratory, form miniature resemblances of those mighty masses, fabricated in the great laboratory of nature? She set to work with characteristic zeal, mixing carbonate of lime, now with one proportion of marine vegetable and animal matter, and then with another, till after numerous failures she at last discovered the right proportions, by which, in the course of a few days or at farthest a few weeks, she can form artificial rocks of the firmest structure. She saw that her discovery could be turned to excellent account in many ways, but chiefly by forming a cheaper and a superior substitute for lath and plaster and ornamental cornices in dwelling-houses. She has taken out a patent for it under the name of *Intonaco*, and, as it harbours no vermin, is impervious to damp, and lessens greatly the risk of destruction by fire, it is beginning, I believe, to yield her a well-merited remuneration.

But though Sea-weed were utterly unserviceable to the human race, would it therefore be useless? Has the Creator no other creatures but thee, vain man? Yes, he has in every sea, living creatures, myriads of myriads of times more numerous than all the men, women, and children, that live and move upon the face of the earth. And does not God care for them? Yes, verily,—"The earth is full of thy riches, O Lord, so is this great and wide sea, wherein are things creeping innumerable, both small and great beasts. There is that Leviathan which thou hast made to play therein. These wait all on thee, that thou mayest give them their food in due season."* Now, though the monsters of the deep live not on Sea-weeds, they live on creatures which in their turn live on those minute animals that fix on Sea-weeds, both as their food and habitation. I was greatly struck with what is said on this subject by the acute and philosophical Mr. Darwin, in his exceedingly interesting journal of the Voyage of the Beagle:—"In all parts of

* "For what purpose the Creator has filled the sea and the rivers with countless myriads of such plants, so that the flora of the deep waters is as extensive as that of dry land, we can only conjecture: the uses to which they are applied by man are, doubtless, of but secondary consideration; and yet they are of no little importance in the manufactures and domestic economy of the human race."—Lindley's 'Vegetable Kingdom.'

the world a rocky and partially-protected shore perhaps supports, in a given space, a greater number of individual animals than any other station. There is one marine production which from its importance is worthy of a particular history; it is the Kelp, or *Macrocystis pyrifera*. This plant grows on every rock from low-water mark to a great depth, both on the outer coast (of Tierra del Fuego) and within the channels. I believe, during the voyages of the Adventure and Beagle, not one rock near the surface was discovered, which was not buoyed by this floating weed. The good service it thus affords to vessels navigating near this stormy land is evident; and it certainly has saved many a one from being drowned. I know few things more surprising than to see this plant growing and flourishing amidst those breakers of the Western Ocean, which no mass of rock, let it be ever so hard, can long resist." Though the stem of this plant is not above an inch in diameter, it is of great strength and surprising longitude. Captain Cooke says that some of it grows to the length of 360 feet and upwards. And yet it is much surpassed in this respect by another, *D'Urvillæa utilis*, I think, which grows to the amazing length of 1500 feet. Mr. Darwin adds:—"The number of living creatures of all orders whose existence intimately depends on the Kelp, is wonderful. A great volume might be written describing the inhabitants of one of these beds of Sea-weed. Almost all the leaves, excepting those that float on the surface, are so thickly encrusted with corallines as to be of a white colour. We find exquisitely delicate structures, some inhabited by simple hydra-like polypi, others by more organized kinds, and beautiful compound *Ascidia*. On the leaves, also, various patelliform shells, Trochi, uncovered mollusks, and some bivalves are attached. Innumerable crustacea frequent every part of the plant. On shaking the great entangled roots, a pile of small fish, shells, cuttle-fish, crabs of all orders, sea-eggs, star-fish, beautiful *Holothuriæ*, *Planariæ*, crawling nereidous animals of a multitude of forms, all fall out together. Often as I recurred to a branch of the Kelp, I never failed to discover animals of new and curious structure. — I can only compare these aquatic forests of the Southern Hemisphere with the terrestrial ones in the intertropical regions. Yet if in any country a forest was destroyed, I do not believe nearly so many species of animals would perish as would here, from the destruction of the kelp. Amidst the leaves of this plant numerous species of fish live, which nowhere else could find food or shelter; with their destruction, the many cormorants and other fishing-birds, the otters, seals, and porpoises, would soon perish also; and lastly the Fuegian savage, the miserable lord of this miserable land, would redouble his cannibal feast, decrease in numbers, and perhaps cease to exist." If such be the amount of animal life and enjoyment on a single bed of Sea-weeds in the Antarctic Ocean, how vast must be the amount of life and enjoyment to which they are necessary in all seas, and how much must they eventually add to the wealth, comfort, and happiness of man?

After all that has been said of the uses of Sea-weeds in agriculture, in medicine, in culinary purposes, in the fine arts, and in various manufactures, I would recommend the study of this department of natural science to my young friends, chiefly on account of the intellectual pleasure it will yield them, and because of its tendency to cherish devotional sentiments in their hearts.

"There is something positively agreeable," says Lord Brougham, "in gaining knowledge for its own sake. There is also a pleasure in seeing the uses to which knowledge may be applied.—It is another gratification to extend our inquiries, and find that the instrument or animal is useful to man, even though we have no chance ourselves of ever benefiting by the information." "But how much more vivid," subjoins Mr. Paterson, of Belfast, "this emotion becomes when we have the pleasure of seeing the beneficial effects of one animal or plant in giving employment to thousands, and multiplying the comforts of the whole civilized world. Is it needful to adduce, as an example, the silk-worm or the cotton-plant?" Great also is the pleasure in studying the physiology of plants, in watching their growth, in examining their structure, in observing the evidences of design in the adaptation of one part to another, and the arrangement of means to an end.

The naturalist knows nothing of that *tædium vitæ*,—that vampire, *ennui*, which renders life a burden to thousands. To him every hour is precious. He may have little leisure for his favourite pursuits; but even those scraps of time which occur in the busiest life, and which many allow to be lost, he gathers up as precious fragments. Habits of observation, of patient research, of accurate discrimination, and orderly arrangement are gradually acquired. Wherever he is—on the wild moor or on the shore of the sea, he learns to see thousands of beautiful, wonderful things which the untrained, uninitiated eye never observes. Is he healthy? His rural rambles are conducive to the continuance of health. Is he in search of health? Health flees from the man who sets out in the direct pursuit of it. But let him have an interest in the wonders of nature—in the works of God's hand,—meditating on them, he forgets his ailments, and health, which he ceases to pursue, by the blessing of God often comes as it were of his own accord. His mind is soothed and refreshed, and the salutary influence is felt by the enfeebled body.

> "There is a pleasure in the pathless woods,
> There is a rapture on the lonely shore,
> There is society where none intrudes,
> By the deep sea, and music in its roar."—*Byron*.

The greatest advantage, however, of this study is, that if rightly prosecuted, it keeps us continually mindful of the presence of God. " These are thy glorious works, Parent of good,—Almighty!" Were we to regard the phenomena of nature with a constant reference to the great Creator, the world, says Paley, would become a temple, and life itself one continued act of adoration.

But let us beware of expecting too much from the study of Natural Science: The book of nature is one of God's books, and it is worthy of Him,—very precious, and fitted to teach us much. But there is another and a better,—the volume of inspiration.' They are from the same Author, and let both be carefully consulted, if we would be wise and happy and good.

Among the advantages arising from the study of Algology, we would mention the great additional enjoyment which it gives to a person who has at any time an hour to spend on the sea-shore, especially if the locality be new to him. We can easily suppose that it must greatly lessen the tedium of a sea voyage, and I am glad to be able to give the following statement from my excellent friend Professor Scouler, of Dublin, who can speak from experience on the subject. "As to my own experience, there is one thing which makes me always bear a kindly regard to Algæ. When at sea for months, the capture of a mass of floating Sea-weed has often given me pleasant occupation for days. Such masses of Algæ may be considered as a marine zoological garden, rich in various animals of every invertebral family. Indeed, the variety of living beings supported in a handful of Sea-weed is truly wonderful. At first sight we select *Serpulæ*, several corallines, such as *Sertulariæ* and *Flustræ*. We observe that interesting mollusk, the *Hyalæa*, climbing branches by means of its beautifully-adapted grooved foot, and grazing upon the fronds. As the animal climbs by means of its abdominal foot with its back often undermost, it has a resemblance so far to the Sloth of the forests of South America. Towards the root of our Sea-weeds we find sponges of small size and curious forms; and here we arrive at a region of still greater activity: we shall probably detect sea-stars and sea-urchins, and amidst them many restless isopode crustaceans; and probably many kinds of articulated worms. Even the sandy matter mingled with the roots is not to be neglected. It must be carefully collected and examined by the microscope, when we shall not fail to discover many beautiful polythalamous shells, and a rich supply of still more strange *Infusoria*."

There is yet another advantage arising from the study of Algology, and indeed of Natural Science in general, which it would be unpardonable to omit. It is of great importance that the young in particular, should be armed against the artifices of those who, by a plausible mixture of facts and fiction, try to sap the foundation of our holy faith, and too often succeed in throwing stumbling-blocks in the way of the unwary. Religion has nothing to fear from facts, but it rejects fiction, and it is well to be able to separate the chaff from the wheat. By their theory of development,— provided you unwittingly swallow all their pretended facts,

5.

—they will trace the progress of a rational creature, from a little almost invisible *monad* floating in the sea, till the monad becomes a monkey, and the monkey a man. And they will tell you that the oak, the monarch of the woods, has arrived at his dignity by almost imperceptible steps, being, some thousands of years ago, only a humble sea-weed in the universal ocean, it may be a *Halidrys*, which signifies Sea-oak, or *D. sinuosa*, which is called the Oak-leaved Sea-weed. If they are less successful now than they once were, it is because Natural Science is now more generally cultivated than when the theory of development was brought forth by Maillet, and fostered by Lamarck. That you may not be imposed upon by their bold assertions and cunning artifices, it is your duty and your interest to study Natural Science, that you may meet and master these deceivers on their own ground. I am sure that I shall be pardoned for giving a quotation from a work recently published by Mr. Hugh Miller, the author of 'Old Red Sandstone,' &c. Had I space to give the whole passage, the reasoning would be found unanswerable:—

"When Maillet first promulgated his hypothesis, many of the departments of Natural History existed as mere regions of fable and romance, and in addressing himself to the Muscadins of Paris in a popular work as wild and amusing as a fairy tale, he could safely take the liberty, and he did take it very freely, of greatly exaggerating the marvellous and adding fresh fictions to the untrue. And in preparing them for his transmutative theory of a marine into a terrestrial vegetation, he set himself, in accordance with his general character, to show that really the transmutation did not amount to much. 'I know you have resided a long time,' his Indian Philosopher is made to say, 'at Marseilles. Now you can bear me witness that the fishermen there daily find in their nets, among their fish, plants of a hundred kinds, with their fruits still upon them, and though these fruits are not so large and so well nourished as those of our earth, yet the species of these plants is in no other respect dubious. They there find clusters of white and black grapes, peach-trees, pear-trees, prune-trees, apple-trees, and all sorts of flowers. When in that city, I saw in the cabinet of a curious gentleman a prodigious number of those sea-productions of different qualities, especially of rose-trees, which had their roses very red when they came out of the sea. I was then presented with a cluster of black sea-grapes. It was at the time of the vintage, and there were two grapes perfectly ripe.' Now all and much more of the same nature addressed to the Parisians of the reign of Louis the Fifteenth, passed, I doubt not, wonderfully well, but it will not do now, when almost every young girl in town and country is a botanist, and works on the Algæ have become popular."

I cannot close these introductory chapters more appropriately than in the eloquent words of Professor Harvey, of Dublin. "If Naturalists too often neglect the true *use* of this knowledge, and rest satisfied with the knowledge itself, the fault and the loss is their own, and must not be charged to science. It is enough for her if she but furnish food which is capable of nourishing the well-directed heart; it is not her province either to cleanse that heart, or to give it power of digestion. For this she must refer her votary to a higher and holier voice; and if she ever speak of looking 'through nature up to nature's God,' she does so with a humble deference to her elder sister, whose province it is to lead the heart to that contemplation. Science and Religion must not be confounded: each has her several path, distinct but not hostile; each in her way is friendly to man; and where both unite they will ever be found to be his best protectors:—the one a 'light to his eyes,' opening to him the mysteries of the material universe;—the other, 'a lamp to his feet,' leading him to the immaterial, and incorruptible, and eternal. The 'eye,' it is true, will grow dim when the light of this world fails; and happy is he who then has 'a lamp' lighted from heaven, and trimmed on earth, to guide him through the hours of darkness. But the eye must not be blamed, because it is not the lamp; nor should science be disdained, because she leaves us far short of just conceptions of the invisible world. Her highest flight is but to the threshold of religion; for what a celebrated writer has said of philosophy generally, is equally applicable to every branch of scientific inquiry. 'In wonder all philosophy began: in wonder it all ends: and admiration fills up the interspace. But the first wonder is the offspring of ignorance; the last is the parent of adoration. The first is the birth-throe of our knowledge, the last is its euthanasy and apotheosis.' "

> " —— If His word once teach us, shoot a ray
> Through all the heart's dark chambers, and reveal
> Truths undiscerned but by that holy light,
> Then all is plain. Philosophy baptized
> In the pure fountain of eternal love,
> Has eyes indeed: and, viewing all she sees
> As meant to indicate a God to man,
> Gives HIM his praise, and forfeits not her own."—*Cowper*.

CHAPTER IX.

A LIST OF THE BRITISH MARINE ALGÆ.

[*According to the systematic arrangement in Professor Harvey's* PHYCOLOGIA BRITANNICA.]

> "The *names* are good, for how, without their aid,
> Is knowledge gained by man, to man conveyed?
> But from that source shall all our pleasure flow?
> Shall all our knowledge be these names to know?
> Then he with memory blest shall bear away
> The palm from GREW, and MIDDLETON, and RAY:
> No! let us rather seek in grove and field,
> What food for wonder, what for use they yield;
> Some just remark from Nature's people bring,
> And some new source of homage to their King."—*Crabbe.*

SPECIES of which the native locality is doubtful are marked with an asterisk (*); those which are doubtful as species and require further examination are marked with a cross (†).

Series 1. MELANOSPERMEÆ.

Fam. 1. FUCEÆ.

I. SARGASSUM.
   * 1. vulgare, *Ag.*
   * 2. bacciferum, *Ag.*

II. CYSTOSEIRA.
   1. ericoides, *Ag.*
   2. granulata, *Ag.*
   * 3. barbata, *Ag.*
   4. fœniculacea, *Grev.*
   5. fibrosa, *Ag.*

III. HALIDRYS.
   1. siliquosa, *Lyngb.*

IV. Pycnophycus.
    1. tuberculatus, *Kg.*
V. Fucus.
    1. vesiculosus, *L.*
    2. ceranoides, *L.*
    3. serratus, *L.*
    4. nodosus, *L.*
    5. Mackaii, *Turn.*
    6. canaliculatus, *L.*
VI. Himanthalia.
    1. lorea, *Lyngb.*

Fam. 2. LAMINARIEÆ.
VII. Alaria.
    1. esculenta, *Grev.*
VIII. Laminaria.
    1. digitata, *Lx.*
    2. bulbosa, *Lx.*
    3. saccharina, *Lx.*
† 4. Phillitis, *Lx.*
    5. fascia, *Ag.*
    6. longicruris.
    7. Cloustoni.

Fam. 3. SPOROCHNOIDEÆ.
IX. Desmarestia.
    1. ligulata, *Lx.*
    2. viridis, *Lx.*
    3. aculeata, *Lx.*
X. Sporochnus.
    1. pedunculatus, *Ag.*

XI. Carpomitra.
    * 1. Cabreræ, *Kg.*
XII. Arthrocladia.
    1. villosa, *Duby.*

Fam. 4. DICTYOTEÆ.
XIII. Cutleria.
    1. multifida, *Grev.*
XIV. Haliseris.
    1. polypodioides, *Ag.*
XV. Padina.
    1. Pavonia, *Lx.*
XVI. Zonaria.
    1. parvula, *Aresch.*
XVII. Dictyota.
    1. dichotoma, *Lx.*
XVIII. Taonia.
    1. atomaria, *Grev.*
XIX. Stilophora.
    1. rhizodes, *J. Ag.*
    2. Lyngbyei, *J. Ag.*
XX. Dictyosiphon.
    1. fœniculaceus, *Grev.*
XXI. Striaria.
    1. attenuata, *Grev.*
XXII. Punctaria.
    1. latifolia, *Grev.*
    2. plantaginea, *Grev.*
    3. tenuissima, *Grev.*

XXIII. ASPEROCOCCUS.
    1. compressus, *Griff*.
    2. Turneri, *Hook*.
    3. echinatus, *Grev*.
XXIV. LITOSIPHON.
    1. pusillus, *Harv*.
    2. Laminariæ, *Harv*.
XXV. CHORDA.
    1. filum, *Lx*.
    2. lomentaria, *Grev*.

Fam. 5. ECTOCARPEÆ.
XXVI. CLADOSTEPHUS.
    1. verticillatus, *Lyngb*.
    2. spongiosus, *Ag*.
XXVII. SPHACELARIA.
    1. filicina, *Ag*.
    . Scoparia, *Lyngb*.
    3. sertularia, *Bonn*.
    4. plumosa, *Lyngb*.
    5. cirrhosa, *Ag*.
    6. fusca, *Ag*.
    7. radicans, *Harv*.
    8. racemosa, *Grev*.
XXVIII. ECTOCARPUS.
    1. littoralis, *Lyngb*.
    2. siliculosus, *Lyngb*.
    3. fasciculatus, *Harv*.
    4. Hincksiæ, *Harv*.
*   5. scorpioides, *Harv*.
*   6. spinescens, *Harv*.

    7. longifructus, *Harv*.
*   8. amphibius, *Harv*.
    9. tomentosus, *Lyngb*.
    10. crinitus, *Carm*.
    11. pusillus, *Griff*.
    12. simplex, *Ag*.
    13. villum, *Harv*.
    14. distortus, *Carm*.
    15. granulosus, *Ag*.
    16. sphærophorus, *Carm*.
    17. brachiatus, *Harv*.
    18. Mertensii, *Ag*.
    19. Landsburghii, *Harv*.
XXIX. MYRIOTRICHIA.
    1. clavæformis, *Harv*.
    2. filiformis, *Harv*.

Fam. 6. CHORDARIEÆ.
XXX. MYRIONEMA.
    1. strangularis, *Grev*.
    2. Leclancherii, *Harv*.
    3. punctiforme, *Harv*.
    4. clavatum, *Harv*.
XXXI. ELACHISTEA.
    1. fucicola, *Fr*.
    2. flaccida, *Fr*.
    3. curta, *Aresch*.
    4. pulvinata, *Kutz*.
       (*attenuata*, Harv.)
    5. stellulata, *Harv*.

88 SYSTEMATIC LIST OF

6. scutulata, *Fr.*
7. velutina, *Fr.*

XXXII. RALFSIA.
  1. verrucosa, *Aresch.*

XXXIII. LEATHESIA.
  1. tuberiformis, *Gray.*
    (*Corynephora marina*, Ag.)
  2. Berkeleyi, *Harv.*

XXXIV. MESOGLOIA.
  1. vermicularis, *Ag.*
  2. virescens, *Carm.*
  3. Griffithsiana, *Grev.*

XXXV. CHORDARIA.
  1. flagelliformis, *Ag.*
  2. divaricata, *Ag.*

Series II. RHODOSPERMEÆ.

-Fam. 7. CERAMIEÆ.

XXXVI. CALLITHAMNION.
  1. plumula, *Lyngb.*
  2. cruciatum, *Ag.*
  3. floccosum, *Ag.*
  4. Turneri, *Ag.*
    (*repens*, Ag.)
  5. pluma, *Ag.*
  6. barbatum, *J. Ag.* (?)
  7. arbuscula, *Lyngb.*
  8. Brodiæi, *Harv.*
  9. tetragonum, *Ag.*
  10. brachiatum, *Bonnem.*
  11. tetricum, *Ag.*
  12. Hookeri, *Ag.*
    (*spinosum*, Harv.)
  13. roseum, *Harv.*
  14. byssoideum, *Arn.*
  15. polyspermum, *Ag.*
    (*Grevillii*, Harv.)

† 16. fasciatum, *Harv.*
17. Borreri, *Ag.*
18. tripinnatum, *Ag.*
* 19. affine, *Harv.*
20. gracillimum, *Ag.*
21. thuyoideum, *Ag.*
22. corymbosum, *Ag.*
  (*versicolor*, Ag.)
23. spongiosum, *Harv.*
24. pedicellatum, *Ag.*
25. floridulum, *Ag.*
26. Rothii, *Lyngb.*
  (*purpureum*, Harv.)
27. mesocarpum, *Carm.*
* 28. sparsum, *Harv.*
29. Daviesii, *Ag.*
  (*secundatum*, Ag.)
  (*lanuginosum*, Lyngb.)

XXXVII. SEIROSPORA.
  1. Griffithsiana, *Harv.*

XXXVIII. WRANGELIA.
  1. multifida, *J. Ag.*

XXXIX. GRIFFITHSIA.
  1. equisetifolia, *Ag.*
  . simplicifilum, *Ag.*
  . barbata, *Ag.*
  . Devoniensis, *Harv.*
  . corallina, *Ag.*
  ⅔/4. secundiflora, *J. Ag.*
  7. setacea, *J. Ag.*

XL. SPYRIDIA.
  1. filamentosa, *Harv.*

XLI. CERAMIUM.
  1. ciliatum, *Ducluz.*
  2. acanthonotum, *Carm.*
  3. echionotum, *J. Ag.*
  4. flabelligerum, *J. Ag.*
  5. nodosum, *Grev. & Harv.*
  6. pellucidum, *Gr. & Harv.*
  7. strictum, *Grev. & Harv.*
  8. gracillimum, *Gr. & Harv.*
  9. diaphanum, *Ag.*
  10. fastigiatum, *Harv.*
  11. Deslongchampsii, *Chauv.*
  12. decurrens, *Grev. & Harv.*
  13. botryocarpum, *Grev.*
  14. rubrum, *Ag.*

XLII. MICROCLADIA.
  1. glandulosa, *Grev.*

XLIII. PTILOTA.
  1. plumosa, *Ag.*
  2. sericea, *Harv.*

Fam. 8. GLOIOCLADIA.

XLIV. CROUANIA.
  1. attenuata, *J. Ag.*

XLV. DUDRESNAIA.
  1. coccinea, *Bon.*
  2. divaricata, *J. Ag.*

XLVI. NEMALION.
  1. multifidum, *J. Ag.*
  2. purpureum, *Harv.*

XLVII. GLOIOSIPHONIA.
  1. capillaris, *Carm.*
  2. ? purpurea, *Harv.*

XLVIII. NACCARIA.
  1. Wigghii, *Endl.*

XLIX. CRUORIA.
  1. pellita, *Fries.*

Fam. 9. NEMASTOMEÆ.

L. IRIDÆA.
  1. edulis, *Bory.*

LI. CATENELLA.
  1. Opuntia, *Grev.*

Fam. 10. SPONGIOCARPEÆ.
LII. POLYIDES.
    1. rotundus, *Grev.*
LIII. FURCELLARIA.
    1. fastigiata, *Grev.*
LIV. GYMNOGONGRUS.
    1. plicatus, *Mart.*
    2. Griffithsiæ, *Mart.*
LV. CHONDRUS.
    1. crispus, *Lx.*
    2. Norvegicus, *Lx.*
LVI. PHYLLOPHORA.
    1. rubens, *Grev.*
    2. Brodiæi, *J. Ag.*
    3. membranifolius, *G. & W.*
    4. palmettoides, *J. Ag.*
LVII. PEYSSONELLA.
    1. Dubyi, *Crouan.*
LVIII. HILDENBRANDTIA.
    1. rubra, *Menegh.*

Fam. 11. GASTROCARPEÆ.
LIX. KALYMENIA.
    1. reniformis, *J. Ag.*
    2. Dubyi, *Harv.*
LX. HALYMENIA.
    1. ligulata, *Ag.*
LXI. GINANNIA.
    1. furcellata, *Mont.*
LXII. DUMONTIA.
    1. filiformis, *Grev.*

Fam. 12. COCCOCARPEÆ.
LXIII. GIGARTINA.
    1. pistillata, *Lx.*
    2. acicularis, *Lx.*
    3. Teedii, *Lx.*
    4. mamillosa, *J. Ag.*
LXIV. GELIDIUM.
    1. corneum, *Lx.*
    * 2. cartilagineum, *Gaillon.*
LXV. GRATELOUPIA.
    1. filicina, *Ag.*

Fam. 13. SPHÆROCOCCOIDEÆ.
LXVI. HYPNEA.
    1. purpurascens, *Harv.*
LXVII. GRACILLARIA.
    1. erecta, *Grev.*
    2. confervoides, *Grev.*
    3. compressa, *Grev.*
    4. multipartita, *J. Ag.*
LXVIII. SPHÆROCOCCUS.
    1. coronopifolius, *Ag.*
LXIX. RHODYMENIA.
    1. bifida, *Grev.*
    2. laciniata, *Grev.*
    3. Palmetta, *Grev.*
    4. membranifolia, *J. Ag.*
    5. cristata, *Grev.*
    6. ciliata, *Grev.*
    7. jubata, *Grev.*

8. palmata, *Grev.*
  (*sobolifera*, Grev.)

Fam. 14. DELESSERIEÆ.
LXX. PLOCAMIUM.
  1. coccinium, *Lyngb.*
LXXI. DELESSERIA.
  1. sanguinea, *Lx.*
  . sinuosa, *Lx.*
  . alata, *Lv.*
  . angustissima, *Griff.*
  . Hypoglossum, *Lx.*
  . ruscifolia, *Lx.*

LXXII. NITOPHYLLUM.
  1. punctatum, *Grev.*
  2. Hilliæ, *Grev.*
  3. Bonnemaisoni, *Grev.*
  4. Gmelini, *Grev.*
  5. laceratum, *Grev.*
  * 6. versicolor, *Harv.*

Fam. 15. CHONDRIEÆ.
LXXIII. BONNEMAISONIA.
  1. asparagoides, *Ag.*
LXXIV. LAURENCIA.
  1. pinnatifida, *Lx.*
  2. cæspitosa, *Lx.*
  3. obtusa, *Lx.*
  4. dasyphylla, *Lx.*
  5. tenuissima, *Lx.*

LXXV. CHRYSIMENIA.
  1. clavellosa, *J. Ag.*
LXXVI. CHYLOCLADIA.
  1. ovalis, *Hook.*
  2. kaliformis, *Hook.*
  * 3. reflexa, *Chauv.*
  4. parvula, *Hook.*
  5. articulata, *Hook.*

Fam. 16. CORALLINEÆ.
LXXVII. CORALLINA.
  1. officinalis, *Linn.*
  2. elongata, *Ell. & Sol.*
  3. squamata, *Ell. & Sol.*
LXXVIII. JANIA.
  1. rubens, *Lx.*
  2. corniculata, *Lx.*
LXXIX. MELOBESIA.
  1. polymorpha, *Linn.*
  2. calcarea, *Ell. & Sol.*
  3. fasciculata, *Lam.*
  4. agariciformis, *Lam.*
  5. licheniformis, *Dne.*
  6. membranacea, *Lx.*
  7. farinosa, *Lx.*
  8. verrucata, *Lx.*
  9. pustulata, *Lx.*

Fam. 17. RHODOMELEÆ.
LXXX. ODONTHALIA.
  1. dentata, *Lyngb.*

LXXXI. Rhodomela.
    1. subfusca, *Ag.*
    2. lycopodioides, *Ag.*

LXXXII. Bostrychia.
    1. scorpioides, *Mont.*

LXXXIII. Rytiphlæa.
    1. pinastroides, *Ag.*
    2. complanata, *Ag.*
    3. thuyoides, *Harv.*
    4. fruticulosa, *Harv.*

LXXXIV. Polysiphonia.
    . parasitica, *Grev.*
    . subulifera, *Harv.*
    . spinulosa, *Grev.*
    . atro-rubescens, *Grev.*
    1. nigrescens, *Grev.*
        (*purpurascens*, Hook.)
        (*atro-purpurea*, Moore.)
        (*affinis*, Moore.)
    6. furcellata, *Harv.*
    7. fastigiata, *Grev.*

*   8. Richardsoni, *Hook.*
    9. Griffithsiana, *Harv.*
* 10. Carmichaeliana, *Harv.*
  11. Brodiæi, *Grev.*
  12. fibrillosa, *Grev.*
  13. violacea, *Grev.*
? 14. variegata, *Ag.*
  15. Grevillii, *Harv.*
  16. fibrata, *Harv.*
* 17. stricta, *Grev.*
  18. pulvinata, *Ag.*
  19. obscura, *Ag.*
  20. formosa, *Suhr.*
  21. urceolata, *Grev.*
  22. elongata, *Grev.*
  23. elongella, *Harv.*
  24. byssoides, *Grev.*

LXXXV. Dasya.
    1. coccinea, *Ag.*
    2. ocellata, *Harv.*
    3. arbuscula, *Ag.*

Series 3.   CHLOROSPERMEÆ.

Fam. 18. SIPHONEÆ.

LXXXVI. Codium.
    1. Bursa, *Ag.*
    2. adhærens, *Ag.*
    3. tomentosum, *Stack.*
    4. amphibium, *Moore.*

LXXXVII. Bryopsis.
    1. plumosa, *Lx.*

2. hypnoides, *Lx.*
LXXXVIII. VAUCHERIA.
  1. submarina, *Berk.*
  2. marina, *Lyngb.*
  3. velutina, *Ag.*

Fam. 19. CONFERVEÆ.
LXXXIX. CLADOPHORA.
  1. Brownii, *Harv.*
  2. pellucida, *Kg.*
  3. rectangularis, *Griff.*
  4. Macallana, *Harv.*
  5. Hutchinsiæ, *Harv.*
  6. diffusa, *Kg.*
  7. nuda, *Harv.*
  8. rupestris, *Kg.*
  9. lætevirens, *Kg.*
  10. flexuosa, *Dillw.*
  11. gracilis, *Griff.*
  12. Rudolphiana, *Kg.*
  13. refracta, *Kg.*
  14. albida, *Huds.*
  15. lanosa, *Kg.*
  16. uncialis, *Harv.*
  17. arcta, *Harv.*
  18. glaucescens, *Griff.*
* 19. falcata, *Harv.*
XC. RHIZOCLONIUM.
  1. riparium, *Kg.*
XCI. CONFERVA.
  1. arenicola, *Berk.*

2. arenosa, *Carm.*
* 3. litorea, *Harv.*
4. Linum, *Roth.*
5. sutoria, *Berk.*
6. tortuosa, *Dillw.*
7. implexa, *Dillw.*
8. melagonium, *W. & M.*
9. ærea, *Dillw.*
10. collabens, *Ag.*
11. bangioides, *Harv.*
12. Youngana, *Dillw.*

Fam. 20. ULVACEÆ.
XCII. PORPHYRA.
  1. laciniata, *Ag.*
  2. vulgaris, *Ag.*
    (*linearis*, Grev.)
  3. miniata, *Ag.*
XCIII. BANGIA.
  1. fusco-purpurea, *Lyngb.*
? 2. ceramicola, *Lyngb.*
  3. ciliaris, *Carm.*
  4. elegans, *Chauv.*
XCIV. ENTEROMORPHA.
  1. Cornucopia, *Carm.*
  2. intestinalis, *Link.*
  3. compressa, *Grev.*
* 4. Linkiana, *Grev.*
  5. erecta, *Hook.*
  6. clathrata, *Grev.*
* 7. Hopkirkii, *M'Calla.*

* 8. ramulosa, *Hook.*
9. percursa, *Hook.*

XCV. ULVA.
1. latissima, *Linn.*
2. lactuca, *Linn.*
3. Linza, *Linn.*

Fam. 21. OSCILLATORIEÆ.

XCIV. RIVULARIA.
1. nitida, *Ag.*
2. applanata, *Carm.*
3. atra, *Roth.*
4. plicata, *Carm.*

XCVII. SCHIZOTHRIX.
1. Cresswellii, *Harv.*

XCVIII. CALOTHRIX.
1. confervicola, *Ag.*
2. luteola, *Grev.*
3. scopulorum, *Ag.*
4. fasciculata, *Ag.*
5. pannosa, *Ag.*
6. hydnoides, *Harv.*
* 7. cæspitula, *Harv.*

XCIX. MICROCOLEUS.
1. marinus, *Harv.*

2. anguinus, *Harv.*

C. LYNGBYA.
1. majuscula, *Harv.*
2. ferruginea, *Ag.*
3. Carmichaelii, *Harv.*
4. flacca, *Harv.*
5. speciosa, *Carm.*

CI. OSCILLATORIA.
1. littoralis, *Carm.*
2. spiralis, *Carm.*

CII. SPIRULINA.
1. tenuissima, *Kg.*

Fam. 22. NOSTOCHINEÆ.

CIII. MONORMIA.
1. intricata, *Berk.*

CIV. SPHÆROZYGA.
1. Carmichaelii, *Harv.*
2. Thwaitesii, *Harv.*
3. Broomei, *Thw.*
4. Berkeleyi, *Thw.*
5. Ralfsii, *Thw.*

CV. SPERMOSIRA.
1. litorea, *Kg.*

POPULAR BRITISH SEA-WEEDS.

"The Sea! the Sea! the open Sea!
The blue, the fresh, the ever free!
Without a mark, without a bound,
It runneth the earth's wide regions round;
It plays with the clouds, it mocks the skies,
Or like a cradled creature lies."—*Barry Cornwall.*

Series I. MELANOSPERMEÆ.

Family I. FUCEÆ.

Genus I. SARGASSUM, *Agardh.*

Generic Character. Frond leaved. Leaves stalked, with a midrib. Air-vessels simple, axillary, stalked. Receptacles small, linear, tuberculated (mostly in axillary characters, or racemes). Seeds in distinct cells.—The generic name is from *Sargazo*, the

Spanish term for masses of Sea-weed found floating in the ocean in some latitudes.—*Greville.*

1. SARGASSUM VULGARE, *Ag.*
2. ———— BACCIFERUM, *Ag.*

Plate III. fig. 4. *S. bacciferum*, with its *berry-like* vesicles, which have sometimes been called *sea-grapes.*

Habitat. Though both of these have been found cast on the shores of the Orkney Islands, and the latter by Mr. W. Backhouse on the English shores, they have no just claim to take rank in our British Flora. But though they came to us only like shipwrecked mariners of another country, who could feel in his heart to cast them out? If we lay hold of them, it is not to treat them roughly, as intrusive aliens, but to give them a kindly welcome, as interesting strangers. We treat them in the same manner as our ornithologists treat a rare and beautiful straggler, which in some of its long migratory flights has been driven by stress of weather to make our island a temporary resting-place.

This wandering Sea-weed has, however, been a very interesting plant to us, and we doubt not to many, since school-boy days, when we read, with all the fascinating charm of novelty, the discovery of a new world by the magnanimous Columbus. " When about 400 leagues to

9.

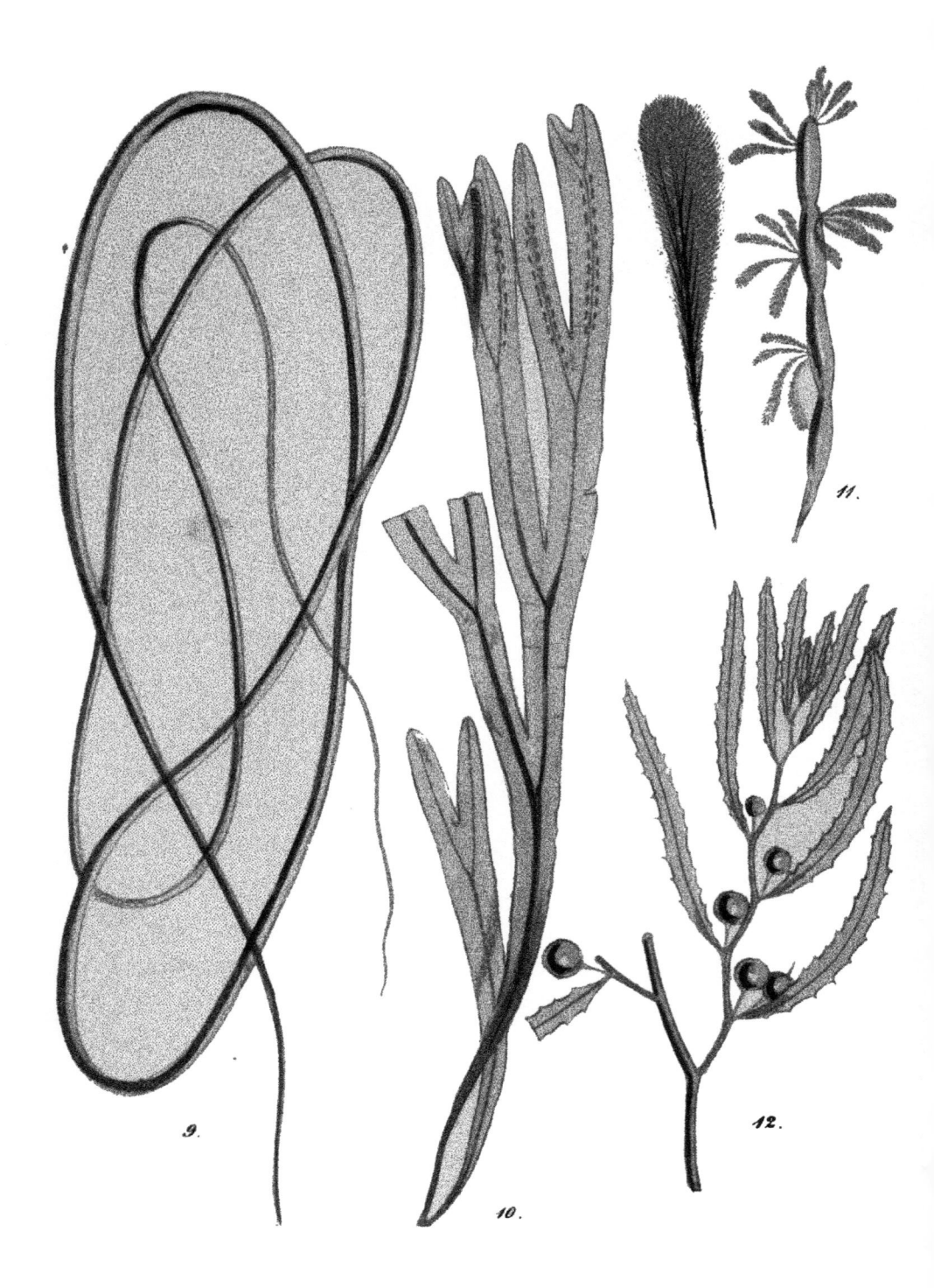

10.

11.

12.

the west of the Canaries, he found the sea so covered with weeds, that it resembled a meadow of vast extent, and in some places they were so thick as to retard the motion of the vessels. This strange appearance occasioned new alarm and disquiet to the sailors. They imagined that they were now arrived at the utmost boundary of the navigable ocean, that these floating weeds would obstruct their further progress, and concealed dangerous rocks, or some large tract of land, which had sunk, they knew not how, in that place. Columbus endeavoured to persuade them that what had alarmed, ought rather to have encouraged them, and was to be considered as a sign of approaching land. At the same time a brisk gale arose, and carried them forward. Several birds were seen hovering about the ship, and directed their flight towards the west. The desponding crew resumed some degree of spirit, and began to entertain fresh hopes."*

Sargassum is found over a wide extent of ocean, but because it was early observed to be very abundant in the Gulf of Mexico, it has very generally been called "gulf-weed." It seldom fails to attract the attention of landsmen, passengers to or from foreign countries; and even sailors, who are less disposed to attend to such appearances than natu-

* Robertson's 'History of America,' vol. i. p. 120.

ralists could wish, not unfrequently bottle up some of the gulf-weed as a curiosity for their friends at home. One might at first think that floating meadows of many miles in extent could be of no service in the middle of the ocean; but they probably support a greater number of living creatures than the richest and most extensive meadows in Britain. They afford both food and shelter to myriads upon myriads of Mollusca, Radiata, Fishes, Crustaceans, &c., many of which are seen playing about, making excursions into the surrounding deep, returning to wanton and hunt amongst the branches, or to rest on them as their home. Many, however, of these living things are parasitical, and attach themselves to the Sea-weeds " for better or for worse." So long as the *Sargassum* floats, they are safe, and are without any effort on their part transported by it from place to place; when the gulf-weed by currents and tempests is wrecked, they perish along with it. When specimens of gulf-weed come in their way, let my young friends scrutinize them, if they admire zoophytes. Not unfrequently have I observed the whole of the berry-like airvessels covered with the finest lace-work, the production of little Polypes, forming a much more delicate *Flustra* than any found in our British seas. I have often seen also the leaves and branches fringed with *Laomedea volubilis,* or something so like our British zoophyte of that name, that without minute examination it could not be distinguished from our own tiny climbing coralline; and at times, though seldom, I have detected a very beautiful *Plumularia,* very like our own no less beautiful *P. cristata,* or Podded Coralline.

Some of my young friends may know that our native coralline, of which I speak, is like a graceful tuft of cream-coloured feathers, and that it is the work of thousands of little active polypes,—a circumstance that renders it still more interesting than our beautiful Sea-weeds. "Each plume," says Lister, "may comprise from 400 to 500 polypi. Many specimens, all united by a common fibre, and all the offshoots of one common parent, are often located on one Sea-weed, the site then of a population which neither London nor Pekin can rival!"*

Though only two species of the gulf-weed have been found on the British shores, the number of species in the genus is very great. In Agardh's 'Synopsis' there are ninety-five species enumerated, and several new and beau-

* See Dr. Johnston's admirable 'History of British Zoophytes,' page 92, plate xxiii. fig. 1–3.

tiful kinds have lately been described and figured by Dr. Greville, in the Annals of Natural History.

The gulf-weed is eaten in China. In the East it is used in salads, and with vinegar it furnishes a pickle.

LICHINEÆ.

LICHINA, *Agardh*.

Generic Character. Frond cartilaginous, blackish-green, dichotomous. Fructification, roundish capsules of the same colour, containing radiating moniliform lines of pellucid seeds imbedded in a gelatinous mass of filaments.—*Greville*.

1. LICHINA PYGMÆA.
2. ——— CONFINIS.

As I have adopted the systematic arrangement given by Professor Harvey in the first volume of his 'Phycologia Britannica,' and, as *Lichina* is not found in it, these two little plants are introduced as interlopers, without any regard to system. Acharius and Sir J. E. Smith ranked the latter one among lichens. Dr. Greville says, "In regard to habit, the *Lichineæ* touch closely on the boundary of the lichens." Beautiful and very instructive figures of both may be seen in Dr. Greville's 'Algæ Britannicæ,' pl. vi., where it may be learned how much the two species differ in fructification, the capsules of *Lichina pygmæa* being subglobose and sessile upon the frond, whereas the capsules of *L. confinis* are oval and terminal. The generic name implies its resemblance to lichens, among which Prof. Harvey, it is probable, means to place it, as he has omitted it in his catologue of Algæ.

To whatever department these little plants may be found to belong, insignificant though they may seem, they are far from being useless. They give variety to the appearance of the otherwise barren-looking rocks on the sea-shore. We are not always sufficiently aware how much our kind Creator consults the happiness of man, even in making what would be offensive, or at least unpleasing, all " beauty to the eye." Look at an old ruinous stone wall by the roadside,—under the shade, it may be, of some overhanging trees. Were it a bare ruin, it would be a disagreeable object; but it is covered with mosses and lichens of all shapes and hues, which so change its aspect that it really gives pleasure even to those who think not how the effect is produced, and know nothing about the mosses and lichens by which the mural ruin is enriched. Now our little *Lichinæ* play their part in effecting this benignant purpose on the rocks upon the shore. Though rather ·lurid in hue, as is also their neighbour *Grimmia maritima,* they form a good contrast with the natural colour of the rocks, and with the grey and yellow lichens with which they are frequently intermingled; and feeble though they may seem, they form some defence to the rocks themselves against the wasting efficacy of the beating surge and the grinding sand; and though many of them may be exhausted by a winter's campaign, by the breath of spring being quickened they return " once more unto the breach."

And then what a snug refuge do these crowded tufts of *Lichinæ* form to innumerable little *Mollusca,* which lurk under them, enjoying at one time the overflowing tide, during which they feast on what the tide brings them; and then, it may be, enjoying little less the security and total ease which fall to their lot during the hours of ebb-tide. Let any person scrape off a handfull of *Lichinæ,* and he will find, on examination, that he has got along with it numerous specimens of *Sphæria depressa, Turtonia minuta,* minute *Littorinæ,* and other Mollusks. We have only to add that *Lichinæ* are found on all our sea-side rocks.

"Roll on, thou deep and dark blue ocean, roll!
Ten thousand fleets sweep over thee in vain.
* * * * *
Thy shores are empires, changed in all save thee.
* * * * *
Time writes no wrinkle on thine azure brow:
Such as creation's dawn beheld, thou rollest now."—*Byron.*

Genus II. CYSTOSEIRA, *Agardh.*

Generic Character. Frond furnished with branch-like leaves, becoming more filiform upwards. Air-vessels simple, arranged consecutively within the substance of the branch-like leaves. Receptacles cylindrical, more or less lanceolate, tuberculated, terminal. Seeds in distinct cells.—The name is from two Greek words signifying a little *sac* and a *chain.*—*Greville.*

1. CYSTOSEIRA ERICOIDES, *Agardh.*

Habitat. Rocks in the sea. Perennial. Summer and autumn. Devonshire, Mrs. Griffiths; Cornwall, Mr. Ralfs; Yarmouth, Miss Turner; Bantry Bay, Miss Hutchins. It was once found by me on the coast of Ayrshire at an early stage of my Algological studies. Its heather-like aspect being new to me, I sent it to Sir W. J. Hooker, who fanned my incipient zeal by marking it *Cystoseira ericoides,* new to Scotland! It was found also by my talented friend, Daniel

Curdie, M.D., on the shore of the island of **Gigha**, off Kintyre. It takes its specific name from its resemblance to heath. It has in a very remarkable degree the property of being iridescent when under water in a growing state. In drying it becomes nearly black, and does not adhere to paper.

I shall merely name the other British species of which I have specimens, but have never seen in a growing state. So far as I known, not one has been found in Scotland.

2. *Cystoseira granulata,* Ag.   4. *Cyst. fœniculacea,* Grev.
* 3. ——— *barbata,* Ag.   5. —— *fibrosa,* Ag.

Genus III. HALIDRYS, *Lyngb.*

Generic Character. Frond compressed, coriaceous, linear, pinnated with distichous branches. Air-vessels lanceolate, stalked, divided by transverse septa. Receptacles lanceolate, stalked, compressed. Seeds in distinct cells.—The name is from two Greek words signifying *sea* and *oak.*—*Greville.*

1. HALIDRYS SILIQUOSA, *Lyngbye.* (Plate I. fig. 2. Portion of frond.)

The frond is narrow, compressed, branches distichous, alternate, vesicles stalked, oblong; receptacles stalked pod-like, and hence the specific name; perennial. In fruit in summer; colour olive; the root is an expanded disc, which attaches itself so firmly to the rocks that it requires a man's strength to pull it off; and when torn off by the strong billows, it frequently brings a scurf of the rock along with it. The disc at the base is often considerably more than an inch in diameter; the stem above this disc is half an inch and upwards. The branches are often four feet in length, and they are numerous and bushy.

Habitat. In pools among the rocks between low and high-water marks. Common on all our shores. Found also in the North Sea and the Northern Atlantic. Something like it must have existed in the ancient world, for in a limestone quarry at Ardrossan I found dark impressions on the rock very like a bushy *Halidrys*.

Halidrys siliquosa is a very common, but is also a very handsome plant. It will be regarded with greater interest by my young friends, when they learn that it is often instrumental in bringing within our reach beautiful zoophytes, which are even more attractive, as we have said, than Sea-weeds, because they are living creatures. They form calcareous habitations, which in many cases resemble little shrubs, and hence the name *Zoophyte*, which is derived from two Greek words, the one signifying a *living creature,* and the other a *plant*; for though they do not vegetate like plants, the habitations formed by the numerous little polypes are in not a few cases plant-like. These attach themselves to rocks and shells and Sea-weeds, and *Halidrys siliquosa* is a special favourite with many of them. Nothing is more common than to find on it rich silvery tufts of *Cellularia reptans.* Occasionally also you may observe on it a zoophyte which is like an elegant tuft of feathers, and which I have already described, viz., *Plumularia cristata,* or Podded Coralline. At other times you may see the branches of *Halidrys* intertwined with another zoophyte of great beauty; this is *Valkeria cuscuta,* taking its specific name from its resemblance to the plant called Dodder. In its collapsed state it is apt to be disregarded by the inexperienced, but when you have seen it spread out on paper, you will not willingly let it slip. I could enumerate as many more that are often parasitical on *Halidrys.* Who would think that on a single bunch of Sea-weed there is so much real enjoyment? Whole colonies of happy living creatures, all rejoicing in life, and showing forth the praise of Him by whom they have been kindly and wonderfully made!

"Look who list thy gazeful eyes to feed
With sight of that is fair; look on the frame
Of this wyde universe, and therein read
The endless kinds of creatures which by name
Thou canst not count, much less their nature's aime,
All which are made with wondrous wise respect,
And all with admirable beauty deckt."
SPENSER: *Hymn on Heavenly Beauty.*

Genus IV. PYCNOPHYCUS, *Kütz.*

Gen. Char. Root composed of branching fibres. Frond cylindrical, dichotomous. Air-vessels, when present, innate, simple. Receptacles terminal, cellular, pierced by numerous pores, which communicate with immersed, spherical conceptacles, containing in the lower part of the receptacles, parietal, simple spores, and in the upper, tufted antheridia.—The name is from two Greek words signifying *thick sea-weed.—Harvey.*

1. PYCNOPHYCUS TUBERCULATUS, *Kütz.*

Hab. In rock-pools. Perennial. Summer and autumn.

This is better known by the name of *Fucus tuberculatus.* Those who have examined it see that it is very different in many respects from *Fucus proper.* The generic name which Kützing has given it, has reference to its structure, being compounded of two Greek words, the one signifying *thick*, and the other a *sea-weed*. It is not uncommon in some parts of Ireland; it is less common in England; and we know not that it has ever been found in Scotland. As I have seen it only in a dried state, I shall not attempt particularly to describe it.

Genus V. FUCUS, *Linn.*

Generic Character. Frond plane, compressed, or cylindrical, linear, dichotomous, coriaceous. Air-vessels, when present, innate in the frond, simple, large. Receptacles terminal (except in *Fucus nodosus*), turgid, containing tubercles imbedded in mucus, and discharging their seeds by conspicuous spores.—*Grev.*

* *Frond flat, with a midrib.*

1. Fucus vesiculosus, *Linn.*
Hab. Common on all the sea-shores.

It is the Sea-ware, Bladder, *Fucus*, Kelp-ware, Black-Tang of Scotland, and sometimes, for what reason I know not, Lady-wrack. In Gothland, according to Linnæus, it is Swine-Tang, because, boiling it, and mixing it with a little coarse dried flour, they give it to their hogs. In the Hebrides, cheeses are dried without salt, being covered with the ashes of this plant, which abounds in salt. In Scania it is used as

13.
14.
15.
16.

thatch and fuel. The root is a large flat disc. The fronds are from two to three feet in length. The air-vessels, as large as nuts, are in pairs; the receptacles, in pairs and often forked, terminate the branches.

There is a variety of this, which is sometimes called *Fucus Balticus*. It is found among grass and moss in marshy ground occasionally overflowed by the tide. It is not attached by roots to anything, and yet, like floating gulf-weed, it grows. In a very pleasant excursion in the island of Arran, in the summer of 1847, with Dr. Greville and Prof. Balfour of Edinburgh, and other friends, Dr. Greville pointed it out to me in abundance on the shore at Brodrock near the quay, on ground saturated with fresh water, but overflowed by high tides. In this state it is very diminutive; but the full-grown plant is exceedingly useful. We have already mentioned that it is employed as winter food for cattle. Lightfoot mentions that during snow-storms in the Highlands, the red deer descends from the wild mountains to the shore to feed on this Sea-weed. He mentions also that Dr. Russell has recommended the saponaceous mucus of the vesicles as very effectual in removing glandular swellings, and says that by calcining the plant in the open air, a black salt powder is procured, having the same medical virtues, and answering well as a dentrifice, by not only removing tartar from the teeth, but also correcting laxity in the gums. We have also lately learned, that, when the vesicles are put for two or three weeks in rum or in spirits of wine, the tincture thus formed has been found very efficacious as an embrocation for removing rheumatic pains.

The great use, however, now-a-days, of this weed along with others, is in the manufacture of kelp and iodine; but of this we have already spoken.

2. FUCUS CERANOIDES, *Linn*. This is sometimes called *Horned Fucus*. It resembles the preceding, but it is much thinner, and more transparent; the midrib is more distinct, and the leafy part is narrower; although it is a more graceful plant than *Fucus vesiculosus*. Some have thought that it is only a variety of *F. vesiculosus* caused by its growing where there is a copious admixture of fresh water; but I have seen it where there was no supply of fresh water.

It is common in many places in Scotland, but is less common in England. It grows on rocks in sheltered bays, and is perennial.

3. FUCUS SERRATUS, *Linn*. *Serrated Sea-weed*. (Plate I. fig. 1. Portion of frond.)

This is very common also on all our shores. It is perennial. The frond differs from the preceding by being serrated. In Scotland it gets the name of black wrack, or prickly tang: it is not so rich in kelp and iodine as the others. It is useful as manure, however. In Norway it is used as food for cattle, mixed with meal. The Dutch use it to cover their crabs and lobsters to keep them alive and moist, preferring it to any other, because it is destitute of that mucus which causes them to ferment and putrefy. It is a handsome species, the fronds on both sides being dotted with pencil-like clusters of whitish capillary fibres, and the fronds being often broad; Dr. Greville has seen them in the Isle of Bute two inches and a half broad.

Like the other *Fuci*, it furnishes hiding-places for mollusks and crustaceans. With certain zoophytes also it is a favourite. The fronds are often partially covered with beautiful lace-work produced by *Flustra*, now called *Membranipora pilosa*, and generally in very finely stellated figures. And still more frequently it is invested with *Sertularia pumila*, the Sea-oak Coralline. The richest specimens I ever saw of *Sertularia pumila* were on this *Fucus* at Leith,—the fronds were quite shaggy with it, and completely covered.

We may mention that both these zoophytes are very

phosphorescent. When roughly shaken in the dark they display brilliant coruscations. Mr. Hassall says: "I lately had an opportunity of beholding this novel and interesting sight to great advantage, when on board one of the Devonshire trawling-boats. The trawl was raised at midnight, and great quantities of corallines were entangled in the meshes of the net-work, all shining like myriads of the brightest diamonds."

> "While thus with pleasing wonder you inspect
> Treasures the vulgar in their scorn reject,
> See as they float along th' entangled weeds,
> Slowly approach, upborne on bladdery beads;
> Wait till they land, and you shall then behold
> The fiery sparks those tangled fronds unfold,
> Myriads of living points; the unaided eye
> Can but the fire, and not the form, descry." *—*Crabbe.*

** *Frond flat, or compressed, without a midrib.*

4. FUCUS NODOSUS, *Linn. Knobbed Wrack.* The root is a large, hard, conical mass, from which spring several branches, from two to four or six feet in length. It is much used in making kelp, though not so productive as some other kinds of wrack. It is called in some places yellow-wrack. In England it goes sometimes by the name

* Vide Dr. Johnstone's 'History of British Zoophytes,' pp. 92, 93.

of sea-whistles, in consequence of the custom which children have of converting the vesicles into whistles. The vesicles serve to buoy up the plant amidst the waves. It is of an olive-green colour; the receptacles are yellow; but the whole plant becomes black in drying, and does not adhere to paper. The air-vessels are called *crackers*; and when cast into the fire, they soon show that they deserve the name by a startling explosion when heated.

Hab. Sea-shores. Common. Perennial. Winter and spring.

5. Fucus Mackaii, *Turn.* This was so named by Mr. Turner in honour of my worthy friend, Dr. J. T. Mackay, of Dublin, whose botanical discoveries have been numerous and valuable. It is found in Connemara, and also in the west of Scotland; but I have never fallen in with it, and it is known to me only by dried specimens.

6. Fucus canaliculatus, *Linn. Channelled Fucus.* This is abundant on rocks on the sea-shore, near high-water mark. Perennial. Summer and winter. "Cattle are exceedingly fond of this plant, and never fail to browze on it in winter, as soon as the tide leaves it within their reach. At this season it is peculiarly wholesome as counteracting the costiveness induced by their ordinary straw-commons." —Carmichael, in Sir W. J. Hooker's British Flora.

Genus VI. HIMANTHALIA, *Lyngbye.*

Generic Character. Frond coriaceous, orbicular, pezizæform. Vesicles none. Receptacles elongated, strap-shaped, compressed, dichotomously divided, springing from the centre of the frond, containing immersed tubercles, furnished with a pore.—*Greville.*

1. HIMANTHALIA LOREA, *Lyngb.* (Plate IV. fig. 13.)

Hab. On rocky shores, common. Annual? Biennial?

High authorities give different answers on this point. Some regard it as annual, as the thongs are produced every year: but others say that the long thongs are only receptacles; that the cup-shaped disc is biennial, and that this part is truly the plant. From what we have seen, we are disposed to agree with the latter. In September 1850, we saw it growing very abundantly towards the close, it would appear, of the first season of its existence at Portstewart, a beautiful place in the north of Ireland. There, at low water, it studded the cliffy rocks like olive-green buttons pretty closely congregated. Dr. Harvey, in the second edition of his excellent Manual, says, "From recent observations, I have no doubt that this plant is biennial." The cup-shaped frond, which adheres firmly to the rock, is more than an inch in diameter. The branches or receptacles with us are not more than six feet in length. When dredging, in August 1849, off the Island of Lismore, in Appin, I saw it growing in such abundance as almost to retard the progress of the boat, for, though well rooted, its floating receptacles covered the surface of the water. Some of them must have been of great length. The one which I took up, without any selection, measured twelve feet: others, I doubt not, were much longer. In Cornwall they are at times even twenty feet long. *Himanthalia* is from two Greek words, of which the English name *sea-thongs* is a translation. The fruit consists of tubercles immersed in the frond, and these tubercles discharge their seeds by pores, which give the *thongs* a spotted appearance. This is remarkably the case, when, after lying on the shore for some time, every pore is covered with a yellow dot, which is the mucus of the plant discharged in the death-struggle which goes on, when, torn from the rocks and tossed out by the waves, it lies withering in the open air. Dr. Neill mentions that in the north of Scotland a kind of sauce for fish or fowl, resembling ketchup, is made from the cup-like or fungus-like fronds of this sea-weed.

Family II. LAMINARIEÆ.

"The water is calm and still below,
For the winds and waves are absent there,
And the sands are bright as the stars, that glow
In the motionless fields of upper air;
There, with its waving blade of green,
The sea-flag streams through the silent water,
And the crimson leaf of the Dulse is seen
To blush like a banner bathed in slaughter."—*Percival.*

Genus VII. ALARIA, *Greville*.

Generic Character. Frond membranaceous, furnished with a percurrent cartilaginous midrib, the stem pinnated with distinct leaflets. Fructification, pyriform seeds, vertically arranged in the incrassated leaflets.—*Greville.*

1. ALARIA ESCULENTA, *Grev. Eatable Fucus.* (Pl. I. fig. 4.)

Hab. Rocky coasts, in deep water, frequent. Annual. Winter and spring.

The name given to it by Dr. Greville is from *Ala*, a wing, from the winged base of the frond. In Scotland, in the Lowlands, it is by some called badder-locks, and hen-ware, which may be a contraction of honey-ware, the name given to it in the Orkney Islands. In some parts of Ireland, Dr. Drummond says that it is called murlins. The portion of it that is eaten is the midrib stripped of the membrane. We have not heard of its being eaten in the west of Scotland. It is a handsome plant, and very tasteful figures of it may be found in Turner's 'Historia Fucorum,' Greville's 'Algæ Britannicæ,' and Harvey's 'Phycologia Britannica.' We have found it in great abundance on the rocky parts of the coast of Ayrshire, the Island of Arran, and at Macrihanish Bay and Southend in Kintyre. In September 1850, we saw it forming at low water a rich fringe on some parts of the Giant's Causeway, on the bold coast of Antrim, and we brought away specimens having the lower part of the stem finely pinnated with the distinct leaflets. Owing to this interesting appendage, it is known at the Causeway under the name of "Purses," because these pinnated leaflets are thought to resemble a Highlandman's purse. It is in fructification about midsummer. In favourable circumstances it grows to a great size, from twelve to twenty feet in length. The best specimens for the herbarium are found in rock-pools. It retains in drying its light olive-green colour, and when young adheres well to paper.

Genus VIII. LAMINARIA, *Lamour.*

Generic Character. Frond stipitate, coriaceous, or membranaceous, flat, undivided, or irregularly cleft, ribless. Fructification, clouded spots of spores, imbedded in the thickened substance of some part of the frond. The name *Laminaria* (*Lam.*) is from *lamina*, a thin plate, in allusion to the flat frond.—*Harvey.*

1. LAMINARIA DIGITATA, *Lamx. Sea-girdles, Tangle, Sea-staff,* or *Sea-wand,* of the Highlanders.

Hab. In the sea, generally in deep water. Perennial. Common.

The root is composed of thick clasping fibres; the stem, which is woody, is from two to six feet in length, and from half an inch to nearly two inches in diameter. It is solid, tough, and in old plants woody, expanding into a frond of from two to six, and occasionally eight feet and upwards in length, and two feet in breadth, deeply cleft into several segments. The colour is olivaceous brown. In its young state it has no woody stem, and the frond is entire, resembling young plants of *Laminaria saccharina*, but thicker and less elegant. The reproduction of the frond in old plants is very curious; but for this I refer to the plate and description in 'Phycologia Britannica.'

In its native state it is well entitled to rank, in Europe at least, among the giants of the marine forest. When a full-grown plant has by stress of weather been torn from its moorings, and stranded on the shore never more to wave in the deep, it is a kind of treasure-trove, a well-stored cabinet of the naturalist. If he is a conchologist, he is almost sure to find the beautiful *Patella pellucida* on the slimy frond; and at the very centre of the stout fibrous roots he is still more sure of *Patella cærulea,* snugly ensconced in a cave which it has dug out for itself, where it is quite safe from everything but an uprooting storm. When the Tangle has come from deep water, it frequently brings with it some rare Nudibranchs,—those elegant little creatures so well figured and described in that admirable monograph with which my friend Mr. Alder and Mr. Hancock are happily engaged. The stem is often adorned with Sea-hair (*Sertularia operculata*); and various kinds of corallines may be found on the roots. After the young naturalist has collected all that can be seen, let him put the tangled root into a basin of sea-water, and in the course of an hour or so he will be astonished to see the Protean tribes of little crabs, annelides, and other strange creeping things that issue forth to reconnoitre their new limited locality. But it is more to our present purpose to mention that the stems are very generally fringed with smaller Algæ, some of which are rare, such as *Delesseria ruscifolia* and *Callithamnion pluma*.

But of what use is this great Alga? Can it be eaten? We have never tasted it, but the young stalks and leaves are eaten along with dulse; and old Gerard tells us that when well boiled, and eaten with butter, pepper, and vinegar, it makes good food. Can the woody stems be turned to good account? To very good account; though we cannot rank high in the list of useful purposes, an amusing one mentioned by Dr. Neill,—that of making knife-handles:— "A pretty thick stem is selected, and cut into pieces about four inches long. Into these, while fresh, are stuck blades of knives, such as gardeners use for pruning and grafting. As the stem dries, it contracts and hardens, closely and firmly embracing the hilt of the blade. In the course of some months the handles become quite firm, and very hard and shrivelled, so that, when tipped with metal, they are hardly to be distinguished from hart's-horn." Neither do we envy the inhabitants of Orkney, Shetland, and the Channel Islands, the use of the plant as fuel. Having abundance of good pit-coal at hand, we are very thankful that we need not have recourse to tangle. Were it converted into peat, we should not be unwilling to use it; and we have seen it thus metamorphosed, but on too small a scale to be useful. This was among sandhills on the coast of Ayrshire, where it had been drifted a considerable way inland by some unusually high tide, and having been deeply covered with driven sand, it had lain, it may be for ages, and had become a layer of peat about two inches thick, in which the stout tubular rind of the tangle-stem in a compressed state was quite distinguishable.

But far from unimportant are the purposes to which it has been put in the formation of kelp, to which the stems, and indeed the whole of this plant, greatly contribute. Who would have thought that burnt Sea-weed would ever have been found useful in the manufacture of such a substance as glass? And yet till lately the materials out of which the best window-glass was formed, were two parts of kelp, and one of fine white sand. The kelp was substituted for the "fossil alkali," which, according to a probable account, was accidentally found to contribute to the formation of glass. According to Pliny, "a merchant-vessel loaded with nitre or fossil alkali having been driven ashore on the coast of Palestine, near the river Bolas, the crew went in search of provisions, and accidentally supported the kettles on which they dressed them, upon pieces of the fossil alkali. The river-sand above which this operation was performed was vitrified by its union with the alkali, and thus produced glass. The important hint thus accidentally obtained was soon adopted, and the art of making glass was gradually improved." Though kelp till lately was chiefly employed in Britain in the manufacture of glass and soap, it is now, as we have already stated, principally manufactured for the iodine it contains, and no Sea-weed is so rich in iodine as this great tangle, especially its woody stems. At a kelp-kiln which we saw in operation at the Giant's Causeway last September, we observed that the kelp-burners were using only the woody stems of *Laminaria digitata*, great bundles of which were spread around ready to be cast into the furnace. The kelp seemed very rich, and we doubt not would yield much iodine.

2. LAMINARIA BULBOSA, *Lamour. Bulbous-rooted Tangle; Sea-furbelows; Furbelowed-hangers.*

Hab. In the sea, in deep water. Perennial.

When in a young state, the frond is plane and undivided, the stem short, with a knob near the root, which is composed of fibres. As the growth of the plant proceeds, the stem becomes flat, and when fully grown it is waved and

17. 18. 19. 20.

curled in a curious manner, which renders it stout, and gives the plant a spring to resist the billows. The knot enlarges and becomes hollow, covering the roots which strike into the clefts of the rock, and fibres proceeding from the bulb strengthen the support, and enable the large plant better to withstand the impulse of the waves. The bulb, being thickly covered over with longish-shaped tubercles, is not unlike a plum-pudding stuck over with cloves or almonds. The bulbs are often cast ashore in winter and spring. The largest bulb with us is about the size of a man's head; and though it would not be sufficiently elegant as the head-gear of a mermaid, a merman might do worse than clap it on his pate as a bonnet, when he raises his head out of the water in a breeze. Some tender little mollusks are wise enough to take up their abode in the hollow of the ball, where all is calm even when the storm is raging outside.

It is strange that this plant escaped the notice of Linnæus, and that even Ray did not distinguish it from *L. digitata*. Mrs. Griffiths has paid particular attention to it, and has measured the size of one which was a sufficient load for a man's shoulders. The bulb was about a foot in diameter. When the fronds were spread out on the ground, they formed a circle of at least twelve feet in diameter. It is the largest of all the European species, and it is precious for all the agricultural and chemical purposes for which *L. digitata* is so useful. It is not rare on the coast of Ayrshire, and we have seen large bulbs cast ashore on the coast of Argyleshire, near the Mull of Kintyre.

3. LAMINARIA SACCHARINA, *Lamour.* Sweet Tangle; Sea-Belt.

Hab. In the sea. Perennial. Very common.

The root is composed of clasping fibres. The stem from an inch to a foot in length. The frond is from a foot to ten feet in length, and from an inch to sixteen inches in breadth. The young plants make fine specimens for the herbarium, keeping their colour, and adhering pretty well to paper. The full-grown plants are not only beautifully waved at the margin like the young plants, but they are frequently bullated and rugose and thickened at the centre. The substance varies from cartilaginous to leathery. The colour is olive-brown, tinged with yellow. It is well deserving of the name of Saccharine, for, as I mentioned before, it has been proved, by my friend Dr. Stenhouse, to be rich in mannite, which is nearly as sweet as sugar. With all this, however, to sweeten it, it is not relished as food; indeed, the Norwegians, we are told, esteem it so lightly that they call it *Toll-tare*, implying that it is fit food for the Fiend. But He who made all things very good, made it for good purposes. It is not despised by the farmer, who finds that it yields nourishment to his crops. It is a great favourite with some of the beautiful "minims of nature," and the young naturalist will find that he is amply repaid for the careful examination of its fronds. Beautiful mollusks may be found gliding along them, and they are the fixed habitation of many zoophytes: *Flustra membranacea* covering it to a great extent with its fine silvery lace-like web; and less spreading zoophytes, such as *Lepralia hyalina* and *Lepralia annulata*, richly dotting it. The latter is considered rare, but it is often found abundantly here on *L. saccharina*, and I have scarcely ever seen it on any other Alga.

> "Huge Ocean shows, within his yellow strand,
> A habitation marvellously planned,
> For life to occupy."

4. LAMINARIA PHYLLITIS, *Lamour.* (Plate XIX. fig. 76.) *Hart's-tongue Laminaria.*

It has been questioned by high authorities whether this be more than a variety of *L. saccharina*. It is more graceful in appearance than the young of *L. saccharina* generally is. It is probable, however, that it is only a handsome little sister of the same family.

5. LAMINARIA FASCIA, *Ag*. This is found in England, Scotland, and Ireland, on rocks and stones near low-water mark. Annual. Summer. Fine specimens of it have lately been found in abundance by Dr. Greville on the Black Rocks at Leith. Professor Harvey, in his 'Phycologia Britannica,' Pl. XLV., includes under it also *L. debilis*, as a broader and limberer variety.

6. LAMINARIA LONGICRURIS.

This species, new to Britain, has lately been found on the shore at Saltcoats by the author; and on the shore, under Dunluce Castle, in the north of Ireland, by Professor Harvey.

At Plate CCLXXXIX. of 'Phycologia Britannica,' after the description of *Laminaria saccharina*, Professor Harvey adds, "A species with a simple frond, and very long stem (*L. longicruris*), in many respects resembling *L. saccharina*, but readily distinguished by the stem becoming hollow, and increasing in diameter upwards, abounds in the Northern Ocean, and should be watched for on the shores of Orkney and Shetland." On reading this, it immediately struck me that this must be the plant which I had got on the Ayrshire coast in November 1849, and which I had described to several algologists. As I had preserved it, I sent it to Dublin to the Professor, telling him that I had got only the stem, which was eight feet long, partly hollow, and laden with *Lepas anatifera* and *Lepas striata*. In the end of August 1850, I was favoured with a letter from him, dated Giant's Causeway, saying that he had not yet seen my specimen; but that he had himself on that day picked up, on the shore under Dunluce Castle, a stem of the unquestionable *L. longicruris*, like mine, without root and without frond, but laden also with barnacles, showing that it had been long adrift, and may have come from Newfoundland or Greenland. Mine, which on his return to Dublin he acknowledged as the true plant, had stood the voyage well, for, though without root and frond, it was as fresh in appearance and in smell as if it had been newly torn from the rock. The stem found by Dr. Harvey was eight feet long, about an inch in diameter at the thickest, and less than a quarter of an inch at the very base. When found again, it may be easily distinguished by being hollow.

7. LAMINARIA CLOUSTONI.

This, as well as the preceding, will soon be figured and described in Phyc. Brit. This Professor Harvey finds on the coast of Antrim, almost as common as the *L. digitata*. He says (*in lit.*), " I think it a good species, or at least a *very decided variety*. It differs from the common *L. digitata* in having a more slender *polished* stem, much compressed above when old,—a much *narrower* lamina, *cuneate* at the base, and *very long* in proportion to the length of the stem,—and the whole plant very dark in colour. In specimens of average size the stem is eighteen to twenty-four inches long, and the lamina of the same fifty-four to sixty inches. It is described in Anderson's 'Guide to the Highlands and Islands of Scotland.'" This, though rarely, is found on the coast of Ayrshire. I have described a specimen of *Flustra membranacea* and *L. digitata* as five feet in length by eight inches in breadth. The plant must have been *L. Cloustoni.*

Family III. SPOROCHNOIDEÆ.

"Je crains que la principale utilité que l'on doit retirer de cette étude se trouve dans les goûts simples qu'elle inspire à ceux qui la cultivent. Le jeune homme qui s'y applique avec ardeur, se dérobe par son moyen aux passions turbulentes du premier âge, et fortifie sans cesse la santé par des exercices agréables."—*Vaucher.*

Genus IX. DESMARESTIA, *Lamour.*

Generic Character. Frond linear, either filiform, compressed, or flat, distichously branched, cellular, traversed by an internal, single-tubed, jointed filament; producing, when young, marginal tufts of byssoid branching fibres. Fructification unknown.—Named in honour of A. G. Desmarest, a celebrated French naturalist.—*Harvey.*

1. DESMARESTIA LIGULATA, *Lamour.* (Plate I. fig. 3.) *Strap-leaved Desmarestia.*

Hab. In the sea; generally in deep water. Annual. Summer. Frequent in the south of England and in the south and west of Ireland; we have got it also in the north of Ireland, at the Giant's Causeway. Not common in Scotland. Orkney, Clouston. Frith of Forth, Lightfoot. We have got it, and know that it is not uncommon at Southend, Kintyre, Argyleshire. It has a wide range:—it is found in Jersey; from the coast of France to the Faroe Islands; and at Cape Horn.

The frond is from two to six feet long; substance at first cartilaginous, becoming flaccid when exposed to the air; colour, when growing, olive-brown, becoming green in the air, and yellowish in the herbarium. The fructification is unknown.

This is a remarkably elegant plant. It was first described by Lightfoot, who gives an excellent figure of it. A good coloured figure of it may be found in Phyc. Brit., Pl. cxv. When young it adheres to paper, but not afterwards. Little pencils of filaments are produced at the axils of the spines when in a young state, but they soon fall off.

2. DESMARESTIA VIRIDIS, *Lamour.*

Hab. In the sea, on rocks and stones, often in shallow water.

This is *Dichloria viridis*, Greville; so named from its singular change of hue when taken out of the water. It seems to be pretty generally diffused in England, Scotland, and Ireland. It is very common on the Ayrshire coast. Colour olivaceous when growing, though in shallow water it often has that foxy colour mentioned by Dr. Drummond. Its appearance is very delicate and beautiful. Dr. Greville mentions that it has no pencil filaments. In a very young state the main branches are beset with fine filaments, but they are not pencilled. It is exceedingly difficult to preserve these fine filaments in preserving specimens. Though they appear beautiful in the water, it requires the utmost care to preserve them when removed from their native element. We have proof that the Rev. W. S. Hore has succeeded in preserving them, by a beautiful specimen he kindly sent us. Dr. Drummond, of Belfast, also has succeeded; and he says that he allows his specimens, when laid out on the paper, to dry without pressure.

3. DESMARESTIA ACULEATA, *Lamour.* *Spined Desmarestia.*
Hab. In the sea. Common. Perennial.

The root is a hard disc. The frond is without a midrib. From one to three feet long. When young, the branches are beset with distinct pencils of fine green filaments; when these fall off they are succeeded by stiff spines, so that it appears quite a different plant. When old, it becomes harsh and woody. "It is scarcely possible to conceive a more beautiful object than this plant waving its young and delicately-feathered fronds in the water." When it has lost its fine green ornamental pencils, the colour is dark olive. Whenever the plants shoot out young branches, they are always clothed with the fine green filaments which, constantly accompanying growth, probably perform the function of leaves. No fructification has yet been observed on any of these three species of *Desmarestia.* It is mentioned as a remarkable thing of these *Sporochnoideæ,* that they change the colour of other Algæ when put into the same basin of water with them, and cause speedy decomposition when they come in contact with them.

The branches are at times adorned with delicate parasitical Algæ, and still more frequently with *Tubulipora serpens* and other zoophytes. On the coast of Ayr the large tufts of it that are cast out upon the shore are often hoary with *Crisia geniculata,* a rare zoophyte, though not showy; somewhat like the very common *Crisia eburnea.*

Genus X. SPOROCHNUS, *Agardh.*

Generic Character. Frond filiform, solid, cellular, the axis more dense. Fructification lateral, crested, stalked; receptacles composed of horizontal, branching filaments, whorled round a central axis, and producing obovate spores.—The name is from two Greek words, signifying a *seed* and *wool,* because tufts of fibres accompany the fructification.—*Harvey.*

1. Sporochnus pedunculatus, *Ag.* (Plate V. fig. 17.)

Hab. On submarine rocks in deep water. Rare. Annual. Summer and autumn. Anglesea, Rev. H. Davis; Exmouth, Mrs. Gulson; Filey, Mrs. Gatty; Killiney, Professor Harvey; Belfast, Mr. W. Thompson. Till lately it had been found in Scotland only by Mr. Hassall at Prestonpans; but I have before me a finely-tufted specimen, dredged by the Rev. Mr. Pollexfen in Kirkwall Bay, Orkney, and several specimens, in different stages of growth, dredged by Major Martin, in Lamlash Bay, Arran. I have learned that beautiful specimens of it were dredged lately in Lamlash Bay by

Mrs. Balfour. Miss Turner and Miss White find it in Jersey.

Stem 6–18 inches long, set with long, filiform, horizontal branches, simple and mostly alternate. The colour yellowish-green, and becoming brownish in age. "Few objects," says Professor Harvey, "are more attractive to the eye of a botanist than a fine frond of this species, as it waves its feathery branches in the water." He adds, that if the use of the dredge were more general, this and many others would probably cease to be thought rare.

Genus XI. CARPOMITRA, *Kütz.*

Generic Character.— Frond linear, dichotomous, flat, and midribbed (or filiform), olivaceous. Fructification, mitriform receptacles terminating the branches, composed of horizontal branching filaments, whorled round a vertical axis, and producing elliptic-oblong seeds.—The name is from two Greek words signifying *mitre-fruit.—Harvey.*

1. CARPOMITRA CABRERÆ, *Kütz.*

This is extremely rare. Found by Miss Ball, in 1813, at Youghal. Plymouth Sound, Rev. W. S. Hore and Dr. Cocks, from both of whom we have had the pleasure of receiving fine specimens. It was discovered at Cadiz, and described by Clemente in his list of Spanish Algæ, and the specific name is in honour of his friend, Don Antonio Cabrera.

Genus XII. ARTHROCLADIA, *Duby*.

Generic Character. Frond filiform, cellular, with an articulated tubular axis, nodose; the nodes producing whorls of delicate, jointed filaments. Fructification, pedicellate moniliform pods, borne on the filaments, and containing at maturity a string of elliptical spores.—The name is from two Greek words, signifying a *joint* and a *branch.*—*Harvey.*

1. ARTHROCLADIA VILLOSA, *Duby*.

Hab. On submarine rocks, shells, &c., and on *Zostera* in deep water. Rare. Annual. Summer and autumn. South of England. Yarmouth, Mr. Turner; Anglesea, Rev. H. Davis; Exmouth, Mrs. Gulson; Frith of Forth, Mr. Hassall; Ardthur, Capt. Carmichael; Cumbraes, Major Martin; Wicklow, Prof. Harvey; Malahide, Mr. M'Calla; Jersey, Miss White and Miss Turner.

This elegant plant was formerly included in the genus *Sporochnus*. It was afterwards made the type of a new genus, under the name of *Arthrocladia*, by Duby in France, and *Elaionema* by a highly distinguished British botanist, the Rev. M. J. Berkeley. The French naturalist had rather the priority. The credit of having first pointed out the podded moniliform fructification is due to our distinguished countryman, the Rev. M. J. Berkeley.

Dr. Greville mentions that Mr. Hassall, the discoverer of it in Scotland, observed that fresh specimens, when spread upon paper, rendered it transparent, as if it had been touched with oil; hence the generic name *Elaionema*, which signifies *oily thread*. *Desmarestia ligulata*, and some other Algæ, have the same property. Like the *Desmarestiæ*, it changes its colour when exposed to the air, and hastens on the decomposition of other delicate Algæ that are put along with it in the same basin. In drying, it adheres well to paper. It is found in the Atlantic, Baltic, and Mediterranean Seas.

Family IV.   DICTYOTEÆ.

"Remote from busy life's bewildered way,
O'er all his heart shall taste and beauty sway;
Free on the sunny slope or winding shore,
With hermit steps to wander, and adore!"
Pleasures of Hope.

Genus XIII.   CUTLERIA, *Grev.*

Generic Character. Root, a mass of woolly filaments. Frond

flat, veinless, somewhat fan-shaped, irregularly cleft. Fructification, minute tufts of capsules pedicellate, containing several distinct granules.—Named *Cutleria* by Dr. Greville in honour of Miss Cutler, a distinguished British Algologist.

1. CUTLERIA MULTIFIDA, *Grev.* (Plate II. fig. 6.)

Hab. On rocks and shells in deep water. Rare. Annual. Summer and winter. Found in England by Mrs. Griffiths, Miss Cutler, Mr. Turner, Mr. Wigg, Mr. Borrer, Rev. Mr. Hore; in Ireland by Miss Hutchins, Miss Ball, Professor Harvey, Mr. M'Calla; in Scotland by Major Martin. I am not aware that it has been found in Scotland, except in Orkney, as mentioned by Professor Harvey in his Manual, and by Major Martin, who dredged it in Lamlash Bay, Arran, in August 1850.

Substance betwixt cartilaginous and membranaceous; colour reddish-olive; crisp when fresh, but soon becoming flaccid. Dr. Greville mentions that the capsules or utricles are very like the little black fungus found on the leaves of rose-bushes in our gardens.

Genus XIV. HALISERIS, *Tozzetti.*

Generic Character. Frond flat, linear, membranaceous, with a midrib. Root, a mass of woolly filaments. Fruit, ovate seeds, forming distinct *sori*, or groups, mostly arranged in longitudinal lines.—Named from two Greek words, signifying the *sea* and *endive*.—*Greville*.

1. HALISERIS POLYPODIOIDES, *Ag.* (Plate III. fig. 10. Portion of the frond with sori along the midrib.)

Hab. In the sea. Biennial? August and October.

It was first figured as British by Stackhouse in his 'Nereis Britannica.' It is rare. Found in England by Mr. Stackhouse, Mr. Winch, and Mrs. Gulson; in Ireland by Professor Harvey, Miss Ball, and Mr. M'Calla. It has not been observed in Scotland. In Jersey it has been found by Miss White and Miss Turner; and is found in the warmer parts of the world, in Europe, Asia, Africa, and America. It is said to have a disagreeable odour when fresh; but nevertheless we should be glad to find it, as it is a handsome plant. Its specific name is from the resemblance of its fructification to that of the fern called *Polypodium*. It is often proliferous at the midrib. Mrs. Griffiths first discovered that it had two kinds of fructification.

"Art's finest pencil could but rudely mock
The rich grey lichens broidered on a rock,
And those gay watery grots he would explore
—Small excavations on a rocky shore,

That seem like fairy baths or mimic wells,
Richly embossed with choicest weed and shells,
— As if her trinkets Nature chose to hide
Where nought invaded but the flowing tide."

Jane Taylor.

Genus XV. PADINA, *Adanson.*

Generic Character. Root coated with woolly fibres. Frond flat, ribless, fan-shaped, marked at regular distances with concentric lines, fringed with articulated filaments; apex involute. Fructification, linear concentric sori, bursting through the epidermis of the frond, containing at maturity numerous obovate utricles or tetraspores, fixed by their base, and containing four sporules.—*Harvey.*

1. PADINA PAVONIA, *Lamour.* (Plate XIX. fig. 2. Portion of the frond.)

Hab. On rocks in shallow pools at half-tide level. Annual. Summer and autumn. Several places in the south of England, as at Torquay by Mrs. Griffiths, and at Exmouth by Mrs. Gulson. Very abundant in the Mediterranean. There is a tradition mentioned by Lightfoot that it was found by Dr. Cargill at Aberdeen. Professor Harvey, after stating the tradition, adds: " But it has not been found in Scotland in modern times, and I fear there has been a mistake; yet it is difficult to imagine what could have been mistaken for it, so different in appearance is it from all other Algæ."

Dr. Greville says, "We have few Algæ more singular or beautiful than this." Dr. Harvey says, "Its general resemblance to the expanded tail of the peacock has been noticed by all authors. When viewed growing under water this resemblance is peculiarly striking, the fringes of capillary fibres which adorn it decomposing the rays of light, and giving rainbow-colours to the surface." See a showy figure of it in 'Phycologia Britannica,' Pl. xci.

Genus XVI. ZONARIA, *Areschoug*.

Generic Character. Root coated with woolly fibres. Frond flat, ribless, fan-shaped, entire or cleft, marked with concentric lines, the cells radiating. Fructification, scattered *sori*.—The name is from the Greek word for a *girdle* or *zone*.

1. ZONARIA PARVULA, *Aresch.*

Hab. Rocks in the sea, on shells and on large Seaweeds. Rare. Annual? Spring and summer. Found by Dr. Greville and Miss Cutler at Sidmouth.

The frond is procumbent, attached to other substances by

whitish fibres. The colour is an olivaceous-green. I do not think that it is rare in the west of Scotland, but as it is rather minute it does not attract notice. The first time I observed it was on a shell dredged near the island of Greater Cumbrae. I did not then know it; but on sending it to Professor Harvey, he told me that it was a rare native, then going by the name of *Padina parvula*. During the months of January and February 1849, I often observed it in a young state on the roots of *Halidrys siliquosa* and *Laminaria digitata*, but more frequently on the former, on the solid disc of the base. In the summer of 1850, I have repeatedly dredged it off the Cumbrae Islands and off Arran. It was generally on *Venerupis decussata*, like a dark epidermis, and not unfrequently the lobes were imbricated. It has been dredged in Lamlash Bay by Professor and Mrs. Balfour. No fructification has yet been observed in Britain; but it has been found on Swedish specimens, and described by Areschoug. The substance is membranous, somewhat transparent, and highly reticulated; the cells quadrangular; colour olivaceous-green. In drying, it does not adhere to paper, and becomes a little darker.

Genus XVII. DICTYOTA, *Lamour*.

"Nihil inutile, nihil vanum, nihil supervacaneum in Naturâ."—*Bacon*.

Generic Character. Frond flat, reticulated, membranaceous, dichotomous, or irregularly cleft (palmato-flabelliform in D. *atomaria*). Root, a mass of woolly filaments. Fructification composed of scattered or variously aggregated, somewhat prominent seeds, on both surfaces of the frond.—The name signifies *network*, in allusion to the reticulations of the frond.—*Greville*.

1. DICTYOTA DICHOTOMA, *Lamour*. (Plate II. fig. 5, the broad variety, and the small figure to the left is var. β.)

Hab. In the sea, on Algæ, and in rock-pools; very common. Annual. Summer and autumn.

This plant is widely distributed over the world. On the Ayrshire coast and in the islands of Arran and Cumbrae, var. β is very common, the frond of which is narrow. I have had very beautiful broad-fronded specimens from the coast of Argyleshire, gathered by Rev. Mr. Lambie at the south end of Kintyre. Both kinds are well represented in Plate CIII. of 'Phycologia Britannica,' but with us the var. β is of a much darker colour.

Genus XVIII. TAONIA, *J. Agardh*.

Generic Character. Roots coated with woolly fibres. Frond flat, ribless, somewhat fan-shaped, irregularly cleft, highly reticulated, marked with concentric lines. Fructification, linear, wavy, concentric, superficial *sori*, on both surfaces of the frond, consisting of naked spores, destitute of filaments.—The name is from the Greek word for a *peacock*.

1. TAONIA ATOMARIA, *Good. and Woodw*.

Hab. On rocks in the sea. Annual. Summer. Found in England by Mrs. Griffiths, Mrs. Fowler, Mr. Wigg, Mr. Borrer, Mr. Dillwyn; in Ireland by Miss Ball; in Frith of Forth by Dr. Greville: very rare. I have never seen a Scotch specimen. There is a splendid figure of it forming Plate I. of 'Phycologia Britannica.'

Genus XIX. STILOPHORA, *J. Agardh*.

Generic Character. Root, a small disc. Frond filiform, solid or tubular, branched. Fructification, convex, wart-like sori scattered over the surface, composed of obovate spores nestling among moniliform, vertical filaments.—The name is from two Greek words signifying a *point*, and *to bear*, in allusion to the dot-like fructification.—*Harvey*.

1. STILOPHORA RHIZODES, *J. Agardh*.

Hab. Near low-water mark on rocks and Algæ. Annual. Summer. Common in the south of England and in some parts of Ireland. Got at Rothesay by the Rev. Gilbert Laing, and dredged in Lamlash Bay by Major Martin; obtained also in Jersey, and in the Baltic Sea and on the Atlantic shores.

This is *Sporochnus rhizodes* of Alg. Brit. and of Harvey's Manual. As it differs from the true *Sporochni* in fructification, it has been made the type of a new genus. The warted fructification is densely dispersed over the whole frond. Substance, when fresh, cartilaginous; when kept, soft and slimy. Colour yellowish-brown.

2. STILOPHORA LYNGBYEI, *J. Agardh.*

This is the var. β of British authors, which has now been ranked as a distinct species. It is got abundantly in dredging in Lamlash Bay, Arran. At times it is got with the frond quite compressed, and about a line in breadth. When newly dredged it is often crisp and rigid, and at other times soft and slimy.

Genus XX. DICTYOSIPHON, *Greville.*

Gen. Char. Frond filiform, tubular, continuous, branched.

Root minutely scutate, naked. Fructification, solitary or aggregated naked spores, scattered irregularly over the surface.—The name is from two Greek words, signifying *net-work* and a *tube*, the frond being tubular and reticulated.—*Greville*.

1. DICTYOSIPHON FŒNICULACEUS, *Grev.*

Hab. In the sea, on *Chorda filum* and other Algæ. Annual. Spring and summer. Anglesea, Dillenius; Cornwall, Hudson; Ireland, Miss Hutchins, Dr. Drummond; Frith of Forth, Dr. Greville. On the Ayrshire coast, very common.

It resembles in appearance *Dichloria* or *Desmarestia viridis*. Within the tube the surface is lined with pellucid oblong cellules. Colour, when young, pale yellow or olive-green; when old, it is of a rusty-brown colour, and the plant is then several feet long, and has a coarse appearance.

Genus XXI. STRIARIA, *Greville*.

Gen. Char. Frond filiform, tubular, continuous, membranaceous, branched. Root naked and scutate. Fructification, groups of roundish seeds forming transverse lines.—The name is from the transverse *striæ*, formed by the lines of fructification.

1. STRIARIA ATTENUATA, *Greville.* (Plate II. fig. 8.)

Hab. Parasitical on the smaller Algæ, generally beyond tide range. Found by Dr. Greville in Bute. The branches

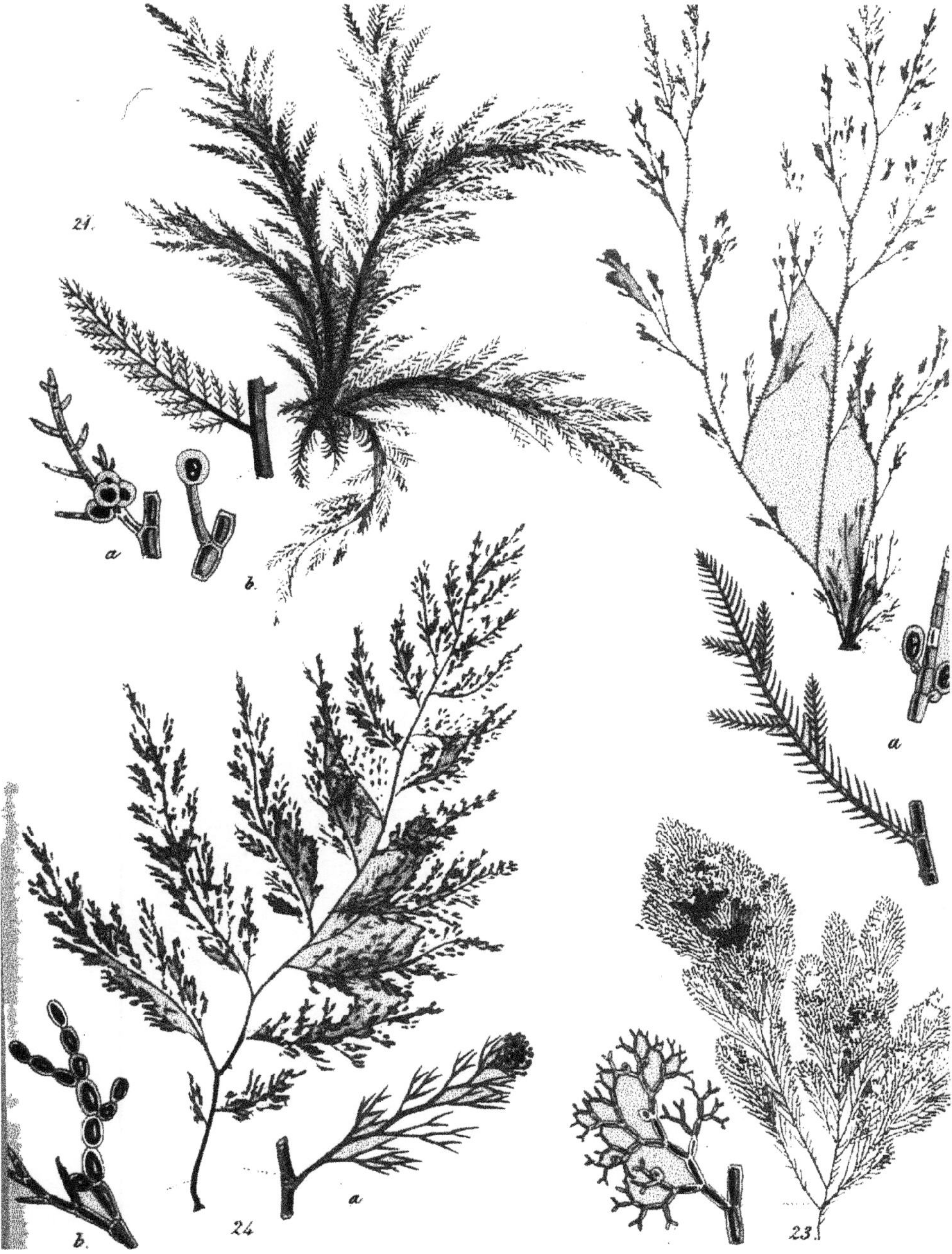
21.
a
b.
a
24
a
b.
23.

are marked, at spaces of half a line asunder, with transverse rings of spores.

Genus XXII. PUNCTARIA, *Greville*.

Gen. Char. Frond simple, membranaceous, flat, with a naked scutate root. Fructification scattered over the whole frond in minute distant spots, composed of roundish prominent seeds, intermixed with club-shaped filaments. — The name is from *punctum*, a dot; the fruit being in dots scattered over the surface of the frond.—*Greville*.

1. PUNCTARIA LATIFOLIA, *Greville*. (Plate IV. fig. 16.)

Hab. On rocks in the sea. Annual. April and May. Found by Mrs. Griffiths at Torquay; Dr. Drummond, near Belfast; Professor Harvey, west of Ireland.

We have found it very large in the island of Arran in the month of June, when the broad pale fronds were partly decomposed. The colour of the frond is pale olive-green, tender, suddenly tapering at the base.

2. PUNCTARIA PLANTAGINEA, *Greville*.

Hab. On rocks and other Algæ. Found by Mr. Turner at Cromer; Dr. Drummond, Belfast; Dr. Greville, Frith of Forth. It is found during summer at Saltcoats.

So late as the month of September I found it cast out on

the shore in great abundance near Brodick in the island of Arran; it was narrow and dark-coloured, and I at first mistook it for *Porphyra linearis* in an old state, wondering that *it* should be found so late in the season: as on the opposite coast of Ayrshire it disappears very early. The fronds in this instance were very much attenuated at the base. The dots of fructification, instead of being round, as in the preceding species, are in this oblong and longer. There is a fine figure of it with magnified fruit, &c., in Alg. Brit., Pl. IX. fig. 2. When this species is fresh gathered it has the perfume of cucumbers so strongly as to fill a room with its fragrance, when the vasculum is opened. Perhaps the other species have the same perfume, but we have not observed it.

3. PUNCTARIA TENUISSIMA, *Greville*.

Hab. In the sea, parasitic on *Zostera marina*. Annual. Summer. Found by Captain Carmichael at Appin; Dr. Greville, in Bute. When in a row-boat off Little Cumbrae we have seen it growing abundantly on *Zostera marina*. It is often got on *Zostera marina* by Major Martin at Ardrossan. We obtained it in June 1850, in great abundance, among the drift on the shore, between Largs and Fairlie, attached to *Zostera*. Mrs. Griffiths thinks that it is the young of *Punctaria latifolia*.

These *Punctariæ* seem to be favourite food of some of the mollusks, as the fronds are often, when found, greatly perforated. Even for the creeping things innumerable in the sea, He who made them provides suitable food. "These wait all upon Thee, that Thou mayest give them their meat in due season. That Thou givest them they gather. Thou openest thy hand, and satisfiest the desire of every living thing."

Genus XXIII. ASPEROCOCCUS, *Lamour*.

Gen. Char. Frond simple, tubular, cylindrical, or (rarely) compressed, continuous, membranaceous. Root minutely scutate, naked. Fructification, distinct spots composed of imbedded seeds, mixed with erect club-shaped filaments. — Name from two words signifying *rough* and *seed.—Greville & Harvey*.

1. Asperococcus compressus, *Greville*. (Pl. V. fig. 18.)

Hab. Parasitical on Algæ in rather deep water. Annual. Summer. Found by Mrs. Griffiths, Torquay; Mr. Ralfs, Mount's Bay; Miss Warren, Falmouth; Miss Turner, Jersey; Cape of Good Hope, Professor Harvey. It has not been found in Scotland.

It was discovered by Mrs. Griffiths at Sidmouth in 1828. Kützing proposes that it should be the type of a new genus, in which case, as Professor Harvey says, *Griffithsianum* may very deservedly be applied. See an excellent figure of it in 'Phycologia Britannica,' Pl. LXXII.

2. ASPEROCOCCUS TURNERI, *Hooker*. (Plate II. fig. 7.)

Hab. In the sea, on stones and the larger Algæ. Found also in rock-pools. Annual. Summer and autumn. Obtained in England in various parts by Mrs. Griffiths, Mrs. Gulson, and Mr. Borrer. In Ireland by Miss Hutchins, Professor Harvey, Mr. Wm. Thompson, Mr. Ball, Miss Ball, Mr. Andrews. In Scotland by Captain Carmichael, Appin; Dr. Greville, Arran. It has been repeatedly dredged off Bute by Mr. W. Gourlie of Glasgow; and it has often been dredged off the islands of Arran and Cumbraes by Major Martin and D. L. In August 1850, specimens fourteen inches long were dredged by Mrs. Balfour in Lamlash Bay, Arran.

The first time I found this in the island of Arran it was in a tide-pool on the rocky shore near Clachland Point. There was a fine large tuft of it, but, being filled with water, it was difficult to distinguish it from the element in which it was growing. The next time, it was found by Dr. Greville, on the same rocky shore, but nearer Brodick. He was glad to lay in a good supply, and three or four more of us who were with him, helped ourselves as liberally, and yet much of it was left in the pool. I looked in vain for it, however, the succeeding season. The specimens were very good, about six inches in length, and nearly an inch in breadth; better fitted for the herbarium than gigantic specimens of it, found in Ireland by Mr. William Thompson, three feet and a half in length, and two inches and a half in diameter! See a very good figure of it with fruit, &c., in Phyc. Brit., Pl. XI.

3. ASPEROCOCCUS ECHINATUS, *Greville.*

Hab. Rocks in the sea. Common. Annual. Summer and autumn.

This is not so beautiful a plant as the preceding. It is much darker in colour, and generally much smaller, at least in diameter. I have, however, seen it betwixt two and three feet in length; this was in the island of Arran, where it had a tomentose appearance in the water, as if the frond were woolly. The same thing seems to have been observed by Captain Carmichael on *Asperococcus* (?) *pusillus,* "beset," he says, "with pellucid fibres so closely covering the frond on which they grow, as to give it the appearance of a bottle-brush." This appearance was very remarkable on the very large specimens which I saw in Arran; but as they were growing in deep water, I could not at the time get hold of them for examination.

Genus XXIV. LITOSIPHON, *Harvey*.

Gen. Char. Frond unbranched, cylindrical, filiform. Fructification, naked spores scattered over the surface. Name from two Greek words signifying *slender tube*.

1. LITOSIPHON PUSILLUS, *Carm*.

Hab. In the sea, parasitical on *Chorda filum* and other Algæ. Annual. Autumn.

This is what was formerly called *Asperococcus* (?) *pusillus*. It has now been made the type of a new genus by Professor Harvey. It is got on the coast of Ayrshire, and is far from being uncommon in the island of Arran.

2. LITOSIPHON LAMINARIÆ, *Harvey*.

Hab. In the sea, parasitical on *Alaria esculenta* and *Ulva lactuca*. It has been found by Drs. Greville and Professor Walker Arnott in the Frith of Forth; by Mr. Ball and Mr. Thompson in Clare; Mr. Moore, Antrim; Capt. Carmichael, Appin.

Genus XXV. CHORDA, *Stackhouse*.

Gen. Char. Root scutate. Frond simple, cylindrical, tubular, its cavity divided by transverse, membranous septa, into separate chambers. Fructification, a stratum of obconical spores, much attenuated at the base, covering the whole external surface of the frond. Among these are found elliptical *antheridia?*—The name signifies a *cord.—Harvey*.

1. CHORDA FILUM, *Lamour*. (Plate III. fig. 9.)

Hab. In the sea, on rocks and stones, very common. Annual? Summer, autumn, and winter.

The structure of this Algæ, which seems cylindrical, is very remarkable, being composed of a fillet spirally twisted into a filiform tube. The colour is olive-green, becoming dark in drying. It is clothed with pellucid hair-like fibres, which, with the mucus of the plant, give it a slippery feel. The length to which it grows in favourable circumstances is very great, even thirty, and at times forty feet. Like *Sargassum*, it forms at some places extensive sea-meadows; but, though floating, it is always under the surface. Dr. Patrick Neill says: "In Orkney we have sailed through meadows of it in a pinnace, not without some difficulty, where the water was between three and four fathoms deep, and where, of course, the waving weed must have been from twenty to thirty feet long. This, too, was the growth of one summer, for the storms of winter completely sweep it from the bay every year." He joins with Lamouroux, however, in thinking that it may not be strictly annual, and that its duration may depend on the nature of the place where it grows. I do not think it is annual, for there is no month, either in winter or spring, when some of it is not floated out on the coast of Ayrshire in stormy weather, and it is often adorned both in winter and spring with a pretty zoophyte, *Laomedea geniculata*, giving it a bottle-brush appearance; the zoophyte is very phosphorescent in the dark. In winter it seems to come from deep water, bringing with it *Millepora polymorpha*, and at times *Venus aurea*, not found on our Ayrshire coast except in this way, though abundant in Loch Ryan.

Lightfoot mentions that the stalks, skinned and twisted when half dry, acquire such toughness as to be used for fishing lines, like Indian grass, which *grass*, Dr. Neill informs us, is an animal substance attached to the ovaries of the small foreign sharks. Something similar is found at the corners of the ovaries of our common dog-fish, by which they anchor themselves to marine plants.

Chorda filum goes by various names. In England it is often called *sea-laces*; in Orkney, *catgut*; in Shetland, *lucky Minny's lines*; in Ayrshire, *dead men's ropes*, and we know an instance in which it proved too deserving of the name: a fine young man, in bathing, being entangled by it and brought out dead. For an excellent figure with the hair-like fibres; for a part of the cylinder untwisted; and for the different kinds of fructification, see Phyc. Brit., Pl. cvii.

2. CHORDA LOMENTARIA, *Greville*.

Hab. Rocks in the sea. Annual. Common.

Dr. Greville describes this well by saying that it is like the intestine of an animal tied at certain intervals. *Asperococcus castaneus*, Br. Flor., is the young of this, and is very common in the spring on the coast of Ayrshire.

Family V. ECTOCARPEÆ.

"Not a tree,
A plant, a leaf, a blossom, but contains
A folio volume. We may read, and read,
And read again, and still find something new;
Something to please, and something to instruct."

Genus XXVI. CLADOSTEPHUS, *Agardh*.

Gen. Char. Fronds inarticulate, rigid, cellular, whorled, with short, jointed, subsimple ramuli. Fructification, elliptical utricles, furnished with a limbus, pedicellate, borne on accessory ramuli. —The name is from two Greek words, signifying a *branch*, and a *crown*.—*Harvey*.

1. CLADOSTEPHUS VERTICILLATUS, *Lyngb*. (Plate IV. fig. 14, a branch of *C. verticillatus*, natural size; and at the base on the left a small portion of a branch with a whorl, magnified.)

Hab. In the sea, on rocks and stones. Perennial. Fruiting in winter. It is pretty common in most places. It is not common, however, on the coast of Ayrshire, but when I say the coast of Ayrshire, I generally mean that part of it with which I have the opportunity of being best acquainted, viz., some dozen miles or so, in the neighbourhood of Saltcoats and Ardrossan. The sea-coast of Ayrshire, strictly speaking, extends nearly ninety miles. On the opposite coast of Arran this species is pretty common.

The filaments are from three to nine inches in length; the colour olive-brown, becoming browner in drying. It does not adhere to paper.

2. CLADOSTEPHUS SPONGIOSUS, *Agardh*.

This is a commoner and a clumsier plant than the preceding.

Genus XXVII. SPHACELARIA, *Lyngbye*.

Gen. Char. Filaments jointed, rigid, distichously pinnated, rarely simple, or subdichotomous. Apices of the branches distended, membranous, containing a dark granular mass. Fructification, elliptical utricles, furnished with a limbus, borne on the ramuli.—The name is from a Greek word signifying a *gangrene*, alluding to the withered tops of the branches.—*Harvey*.

Sphacelariæ are divided into two classes, viz.,

* Those whose fronds are beset with woolly fibres at the base or lower part.

** Those whose stems are naked at the base without woolly fibres.

* *Stems clothed at the base.*

1. SPHACELARIA FILICINA, *Agardh*.

Hab. On rocks and Nullipores near low-water mark, and on the roots of *Laminariæ*, &c. Very rare. Found in England by Mrs. Griffiths, Mrs. Hare, Mrs. Wyatt, Mr. Sconce, Mr. Ralfs, Dr. W. Arnott, Mr. Borrer, and very fine by Mrs. Gulson at Littleham Cove; in Ireland by Miss Ball and Mr. W. Thompson, and Dr. R. Harvey; and in Jersey by Miss Turner and Miss White. Not found in Scotland.

From two to four inches high. Pinnæ alternate;

colour greenish-olive; substance rigid. Professor Harvey says, "There are few more beautiful plants among the filiform Algæ of our coasts, and not many more rare." It is a species of the south of Europe, where it is found much larger than in Britain. Miss Turner, however, has found in Jersey a specimen fully equal in size to those of the south of Europe. (See a fine figure in Phyc. Brit., Pl. CXLII.)

2. SPHACELARIA SERTULARIA, *Bonnem.*

This, though more diminutive, is even more beautiful than the last. It is finely figured in Phyc. Brit., Pl. CXLIII., though Professor Harvey is scarcely disposed to allow that it is any more than a deep-water variety of *Sphacelaria filicina*.

3. SPHACELARIA SCOPARIA, *Lyngbye*. (Plate IV. fig. 15, in its summer state, natural size; the figure to the right is a branchlet, magnified.)

Hab. Submerged rocks, and in tide-pools.

In its summer state, it is a fine bushy *broom-like* plant, as the specific name implies. In its winter state it is so bared of its shaggy branchlets that it might well be taken for another plant. The difference of appearance is well represented in Pl. XXXVII. of Phyc. Brit. Dr. Greville finds it in the Frith of Forth. It is not common on the Ayrshire coast, but found in several places in the Island of Arran.

* * *Stems naked at the base.*

4. SPHACELARIA PLUMOSA, *Lyngbye.*

Hab. On rocks at low-water mark, and in rock-pools. Perennial. Found by Mr. Borrer at Beachy Head; Sir J. Richardson and Dr. Greville, Frith of Forth; by Mr. Ralfs in England and Wales; by Mr. W. Thompson, Belfast Bay; Miss Gower, Howth; Rev. Mr. Pollexfen, Orkney; Major Martin, Ayrshire; D. L., Island of Arran; D. L., jun., Joppa, Frith of Forth.

Professor Harvey says that this beautiful plant is peculiarly a northern one. The figure given in Phyc. Brit., Pl. LXXXVII., is taken from a Welsh specimen, and, though true, I doubt not, to nature, is greatly inferior to those we get in the west of Scotland. Those found at Ardrossan and in Arran are finer in colour and broader in the frond. However, D. L., jun., sent me specimens gathered by him in a rock-pool at Joppa, near Edinburgh, which, though very different from those got by us on the west coast, will stand a comparison with them in beauty. The fronds are numerous; they are scarcely so large as the western ones, but they are beautifully feathered; and instead of being light olive, they are so dark as to be almost black: while the Ayrshire ones were light olive, and almost as broad as the feathers of a robin's wing; those from the Frith of Forth were like the greenish-black of a starling's wing. In some specimens the greenness was increased by *Striatella arcuata*, as a parasite.

5. SPHACELARIA CIRRHOSA, *Lyngbye*.

This is a very common kind, parasitical on larger Algæ. It is very variable in appearance. The most common kind, however, may be seen about the end of summer, detached from the other Algæ, and floating in great abundance, like little round balls. It does not adhere well to paper.

6. SPHACELARIA FUSCA, *Agardh*.

This is a rare species, found in the south of England, and in Wales; it is beautifully figured in Phyc. Brit., Pl. CXLIX.

7. SPHACELARIA RADICANS, *Harvey*.

Rather rare, found in England and Ireland.

8. SPHACELARIA RACEMOSA, *Greville*.

This is allied to the preceding, but is larger.

Genus XXVIII. ECTOCARPUS, *Lyngbye*.

Gen. Char. Filaments capillary, jointed, olive or brown,

flaccid, single-tubed. Fruit, either spherical or lanceolate capsules, borne on the ramuli, or imbedded in their substance.— The name is from two Greek words, meaning *external fruit*.— Harvey.

They are divided into two classes:—
* Secondary branches alternate, flaccid, or secund.
** Secondary branches and ramuli opposite.
　　* *Secondary branches alternate.*

1. Ectocarpus littoralis, *Lyngbye.*

Hab. In the sea, and parasitical on any Alga that comes in its way, preferring, however, the *Fuci* and *Laminariæ*. It is very common.

The shaggy tufts are from 6–12 inches long. It is of a brownish-olive colour, and often rust-coloured, in which case it stains the paper to which it adheres. If any prefer a good green to its natural colour, they have only to dip it for a moment in boiling water, and it comes out a pleasant grass-green. This may please the eye, but it is apt to lead the naturalist astray, as it then approaches nearer to *E. siliculosus*: though even in its scalded state it is coarser and more robust.

2. Ectocarpus siliculosus, *Lyngbye.* (Plate V. fig. 19, natural size; on the left, a branchlet magnified.)

Hab. On Algæ, &c. Common.

It is finer than the last, though sometimes larger. The colour is generally a pale olive, sometimes greenish, and sometimes yellowish. It adheres well to paper, and makes a beautiful specimen. It is best distinguished from the preceding by the fruit, which is podded (hence the specific name), and on short stalks, whereas the fruit of the former is imbedded in the branches, and subglobose.

3. ECTOCARPUS FASCICULATUS, *Harvey*.

This is rather a rare plant.

4. ECTOCARPUS HINCKSIÆ, *Harvey*.

This is rare also. Found by Miss Hinks at Ballycastle, and named in honour of her. We have only once found it in Ayrshire.

5. ECTOCARPUS TOMENTOSUS, *Lyngbye*.

Hab. On rocks and Algæ. Not uncommon.

From one to eight inches long; frond sponge-like; colour sometimes brownish, and at other times a greenish-olive.

6. ECTOCARPUS LONGIFRUCTUS, *Harvey*.

Hab. Skail, Orkney, Mrs. Moffat.

Resembling *E. littoralis*, but the fruit is larger. We got one specimen at Saltcoats, with fruit corresponding to the figure in 'Phycologia Britannica.'

7. ECTOCARPUS DISTORTUS, *Carmichael*.

Hab. Appin, Captain Carmichael. Dredged at Appin in August 1849, by D. L. The filaments are bent in a zigzag manner, and beset with spine-like divaricated ramuli.

Having dredged in Lamlash Bay something that puzzled me, I sent it to Professor Harvey, who at first thought that it was *Ectocarpus distortus* of Carmichael; but having afterwards dredged it himself in Roundstone Bay, he wrote to me that he thought it a different plant from Captain Carmichael's, and, as I had first found it, he named it—

8. ECTOCARPUS LANDSBURGII, *Harvey*.

It has not much beauty to recommend it, but it is a little curiosity. Like the Scotch thistle, it is armed at all points, and says as plainly as a hundred dirks can say it, "*Wha daur meddle wi' me?*" Phyc. Brit., Pl. CCXXXIII. It was dredged a second time by D. L. in Lamlash Bay, in August 1850, but it seems rather rare.

9. ECTOCARPUS CRINITUS, *Carmichael*.

Hab. Muddy sea-shores, "spreading over the mud in extensive fleeces of a bright bay-colour." Found by Capt. Carmichael at Appin, and by Mrs. Griffiths in Devonshire.

10. ECTOCARPUS PUSILLUS, *Griffiths*.

Found by Mrs. Griffiths, Torquay; and by Mr. Ralfs,

Land's End. Of this I have fine specimens from Mrs. Griffiths.

* Secondary branches opposite.

11. ECTOCARPUS GRANULOSUS, *Agardh.*

Hab. In rock-pools on other Algæ. Not uncommon in England and Ireland, and I find that it is pretty common on the coast of Ayrshire.

It is a handsome plant, as may be seen by Pl. cc. in 'Phycologia Britannica.' It differs from its British congeners, by bearing dark-coloured elliptical capsules or *utricles* on the upper side of the opposite branches or branchlets. The fruit is often very abundant, and is quite visible by the naked eye.

12. ECTOCARPUS SPHÆROPHORUS, *Carmichael.*

Hab. On *Ptilota sericea* or *Cladophora rupestris.* This has been found in England, Scotland, and Ireland, by Capt. Carmichael, Mrs. Griffiths, Mr. Ralfs, and Miss Hutchins.

It is not a common plant, and where it is found, Mr. Ralfs observes, that "it is not diffused through the bay, but is confined to the space of a few rocks, on which it forms, as it were, a colony, or is gregarious." It seems also to confine itself to *P. sericea* and *C. rupestris.* See a fine figure of it in Phyc. Brit., Pl. cxxvi., in which may be seen also the globose fruit, either in pairs opposite to each other, or opposite to a branchlet.

13. ECTOCARPUS BRACHIATUS, *Harvey*.

Hab. On *Rhodomenia palmata*. It has been found by Sir W. J. Hooker and Mrs. Griffiths in England; by Miss Ball and Mr. Wm. Thompson in Ireland; and by Miss M'Leish and D. L. on the coast of Ayrshire. It is rare, however. The Ayrshire habitat is at Seamill, some miles north of Ardrossan.

By the fine figure of it in Pl. iv. of Phyc. Brit., it may be seen that the fruit is imbedded in the stem where two opposite branchlets meet.

14. ECTOCARPUS MERTENSII, *Agardh*.

Hab. On mud-covered rocks near low-water mark. Annual. April and May. Rare.

This fine plant, though found by most of our distinguished naturalists in England and Ireland, is rare in Scotland, where as yet it has been found only by the Rev. Mr. Pollexfen in Orkney, and by D. L. jun., at Joppa, on the east coast, and at Saltcoats on the west coast.

It was named by Mr. Turner in compliment to Professor Mertens of Bremen, a distinguished Algologist. It has a fine feathery appearance. The colour is a clear olive, and it bears its fruit on the opposite ramuli. See the beautiful figure of it in Phyc. Brit., Pl. cxxxii.

Genus XXIX. MYRIOTRICHIA, *Harvey*.

Gen. Char. Primary filaments olivaceous, flaccid (simple), beset on every side with simple, spine-like ramuli, which bear from their tips colourless, dichotomous, long-jointed fibres. Fructification, ovate capsules, containing a dark mass of seeds.—The name is from two Greek words, signifying a *thousand hairs*, from the innumerable hair-like fibres which spring from the ramuli.—*Harvey*.

1. MYRIOTRICHIA CLAVÆFORMIS, *Harvey*. (Plate III. fig. 11, plant, natural size, on *Chorda lomentaria*, and on the left a frond magnified.)

Hab. This is found parasitical on *Chorda lomentaria*. It is got in England and Ireland; and has been gathered at Ballantrae, in Ayrshire, by Mr. W. Thompson, who lets nothing escape his observant eye.

2. MYRIOTRICHIA FILIFORMIS, *Harvey*.

Hab. Also parasitical on *Chorda lomentaria*. It is not uncommon in England and Ireland; and on the coast of Ayrshire, and in the island of Arran.

By comparing our figure of the preceding with the figure of *M. filiformis* in Phyc. Brit., Pl. CLVI., we see that the ramuli in the former regularly increase in length from the base, so as to give it a club-shaped appearance, while in this they are nearly of the same length, and are collected in oblong clusters, leaving bare spaces.

Family VI. CHORDARIEÆ.

"Rerum Natura tota est nusquam magis quam in minimis."

On this family I shall not dwell long; not because they are devoid of interest, but because I hasten on to other families whose beauty is more evident and attractive.

Genus XXX. MYRIONEMA, *Greville*.

Gen. Char. Mass gelatinous, (exceedingly minute,) effused, composed of very short, clavate, erect, mostly simple filaments, "fixed at their base and at their expansion." Fruit, capsules at the base among the filaments.—Name from two Greek words signifying *ten thousand filaments.—Greville.*

1. MYRIONEMA STRANGULANS, *Greville*.

Hab. In the sea, parasitical on several *Ulvæ*. Forming

dark brown spots, and, when on *Enteromorpha*, forming a ring round it. I was unacquainted with this till it was pointed out to me by Dr. Greville in the island of Arran, growing on *Enteromorpha* in a rock-pool.

2. Myrionema punctiforme, *Harvey*.

On *Chylocladia clavellosa* at Appin; on *Ceramium rubens*, Mrs. Griffiths, Torquay. A beautiful microscopic object.

3. Myrionema clavatum, *Harvey*.

A thin purplish crust, covering the pebbles at half-tide level, requiring the microscope to detect it.

4. Myrionema Leclancherii, *Harvey*.

The figures of this species and of *M. punctiforme*, in Pl. XLI. of Phyc. Brit., give a better idea of them than any verbal description.

Genus XXXI. ELACHISTEA, *Duby*.

Gen. Char. Parasites composed of simple, vertical, or radiating, jointed filaments, issuing from beneath the surface-cellules of other Algæ: the lower part of the filaments hyaline, and compacted together into a tubercle; the upper half coloured (olive), free. Spores oblong, mostly stalked, affixed to the tubercular base.—The name seemingly from a Greek word, signifying the least.—*Harvey*.

25.

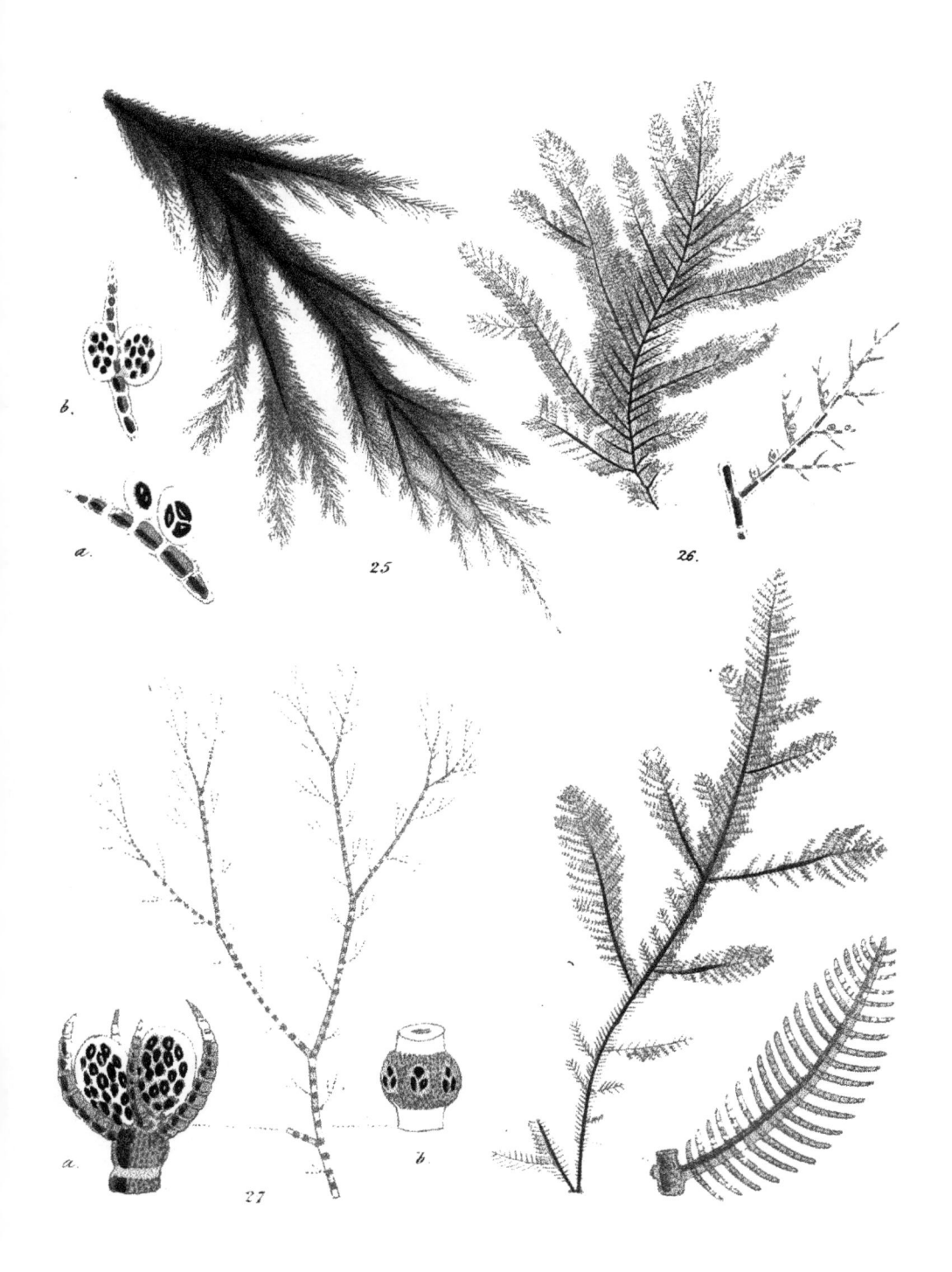

26.

27.

1. ELACHISTEA FUCICOLA (*Conferva fucicola*), *Fr.*
This and the following were formerly placed among *Confervæ*. This one is very common on *Fucus nodosus* and *F. vesiculosus,* forming olivaceous tufts.

2. ELACHISTEA FLACCIDA, *Fr.* Parasitic on *Fucus* and *Cystoseira.*

3. ELACHISTEA CURTA, *Aresch.* On *Fuci.*

4. ELACHISTEA PULVINATA, *Kütz.* On *Cystoseira ericoides.*

5. ELACHISTEA VELUTINA, *Fries.* Parasitical on *Himanthalia lorea.*

6. ELACHISTEA STELLULATA, *Harvey.* On *Dictyota dichotoma.*

This very rare parasitical plant, forming tufts half a line in diameter, resembling minute stars, was, in August 1850, found in great abundance by Professor Balfour of Edinburgh and myself, when dredging in Lamlash Bay. It was found at the same time by Major Martin. We got numerous plants of *Dictyota dichotoma* studded with it. As it had never been found before, except by Mrs. Griffiths at Torquay, I was not sure that it was really this rare plant till I sent it to Professor Harvey.

7. ELACHISTEA SCUTULATA, *Fries.* On *Himanthalia lorea.*

Genus XXXII. RALFSIA, *Berkeley*.

Gen. Char. Frond coriaceo-crustaceous, fixed by its inferior surface, orbicular, concentrically zoned; composed of densely packed, vertical, simple filaments. Fructification, depressed warts, scattered over the upper surface, containing obovate spores fixed to the bases of vertical filaments.—*Ralfsia*, in honour of John Ralfs, Esq., of Penzance, a most acute botanist, whose discoveries among the minute Algæ, especially the *Diatomaceæ*, have thrown great light on that little-known branch of botany.—*Harvey*.

1. RALFSIA VERRUCOSA, *Areschoug*.

Hab. Common on the rocky shores of the British Islands. Perennial. Winter. Though very common on the Ayrshire coast, and still more so in Arran, it is not generally known, as it attracts not notice by its beauty.

Genus XXXIII. LEATHESIA, *Gray*.

Gen. Char. Frond globose or lobed, fleshy, composed of jointed, colourless, dichotomous filaments, issuing from a central point; their spines, which constitute a fleshy coating to the frond, coloured and tufted. Fructification, oval spores attached to the colourless tips of the filaments.—Named *Leathesia*, in honour of the Rev. Mr. Leathes, a British naturalist.—*Harvey.*

1. LEATHESIA BERKELEYI, *Harvey.*

We shall briefly say of this, in the words of Professor Harvey—" a small plant, more curious than beautiful, first noticed by the Rev. M. J. Berkeley, on rocks at Torquay."

2. LEATHESIA TUBERIFORMIS, *Gray.*

This differs from the former, which is a dense and solid substance, by being at first flocculent within, and then hollow. With this species I had long been familiar, without knowing its name, till it was pointed out to me in the island of Arran, by Dr. Greville, as the *Corynephora marina* of Agardh. It is of a light yellow colour, and after a breeze in summer it may be seen in heaps in the little bays, not unlike bunches of hops, were it not for irregularity in size.

Genus XXXIV. MESOGLOIA, *Agardh.*

Gen. Char. Frond filiform, much branched, gelatinous. Axis composed of loosely packed, longitudinal, interlaced filaments, invested with gelatine; the periphery of radiating filaments, whose apices produce clusters of club-shaped, moniliform fibres. Fructification, obovate spores, seated among the apical fibres.—

Named *Mesogloia*, from two Greek words, signifying *viscid* and *middle*, from the gelatinous axis.—*Harvey.*

0. Mesogloia vermicularis, *Agardh.*

Hab. On rocks in the sea. Annual. Summer. Common. No beauty; though the fruit is attractive, as represented in Pl. xxxi. of 'Phycologia Britannica.'

2. Mesogloia virescens, *Carmichael.* (Plate V. fig. 20, a frond of the natural size; on the left, a small portion of the frond magnified.)

Hab. On rocks, stones, and Algæ. Annual. Summer. Common. Found in England by Mrs. Griffiths; in Ireland by Mr. W. Thompson and Mr. M'Calla. It is very common in the west of Scotland, much more so than the preceding, and it is much more handsome.

Even to the naked eye, when skilfully spread out, it is beautiful; the colour being a sweet yellowish-green, and its appearance being villous owing to the length of the filaments, which are set in a loose gelatine. The appearance of a branch under the microscope is singularly interesting.

3. Mesogloia Griffithsiana, *Greville.*

This, which is said to be of a much firmer and more compact substance than *M. virescens*, I have not seen.

Genus XXXV. CHORDARIA, *Agardh*.

Gen. Char. Filiform, much branched, cartilaginous, solid. Axis composed of densely packed, longitudinal, interlaced, cylindrical filaments; the periphery of simple, club-shaped, horizontal, whorled spores, seated among the filaments, and long byssoid, gelatinous fibres. Fructification, obovate spores, seated among the filaments of the periphery.—*Chordaria*, from the Latin word signifying a *cord.—Harvey.*

1. CHORDARIA FLAGELLIFORMIS, *Agardh*.

Hab. Attached to rocks and stones in the sea. Annual. Summer. Common on all our British shores. In some places it grows to the length of three feet, but it is seldom found more than the third of that length in the west of Scotland. It is not thicker than small twine, and hence is generally known by the name of *whip-cord*. When in the water it is seen to be thickly set with very fine fibres, giving it a whitish appearance, and these, along with the mucus, render it very slimy to the touch, though the dark-coloured filaments themselves are firm. In drying it generally stains the paper of a rusty colour. With us it is often, when young, clothed with *Gomphonema paradoxum,* which gives it a beautiful appearance.

2. CHORDARIA DIVARICATA, *Agardh.*

This species was known only as an inhabitant of the Baltic Sea, till it was found by Mr. M'Calla, in October 1845, thrown up in great abundance from deep water at Carrickfergus. It is lighter in colour than the preceding, and the axils are very patent.

"Il y a dans chaque plante bien examinée une preuve vivante de l'existence du grande Etre qui gouverne cet univers. Les divers arrangements que présentent les organes sont autant de petits problêmes proposés par la grande Intelligence à notre faible intelligence, qui en dérive. J'avoue, au moins, pour moi-même, que je n'examine pas une simple fleur sans être étonné de la sagesse qui en a disposée les diverses parties, et sans apercevoir, dans le détail ou dans l'ensemble, le texte de méditations les plus profondes."—*Vaucher.*

Series II. RHODOSPERMEÆ.

"Call us not weeds,—we are flowers of the sea,
 For lovely, and bright, and gay-tinted are we;
 And quite independent of culture or showers:
 Then call us not weeds,—we are Ocean's gay flowers." *

Family VII. CERAMIEÆ.

We have now come to a most interesting portion of our little work, in which we have to treat of the *Florideæ*, so

* Motto on the title-page of a pretty album, called 'Treasures of the Deep,' containing about fifty specimens of Scottish Algæ, prepared by the author's

attractive by the loveliness of their hues, the delicacy of their structure and substance, and the grace and elegance of their forms. When well prepared and placed in an album, they are often taken for exquisite paintings. And no wonder, for

> "Who can paint
> Like Nature? Can Imagination boast,
> Amid its gay creation, hues like hers?
> Or can it mix them with that matchless skill,
> And lose them in each other, as appears
> In every bud that blows?"

These hidden beauties would almost make us wish that we could occasionally take a morning walk in the rocky submarine valleys where they grow; or that we could not only call, but bring up from the "vasty deep," some of these splendid treasures. Had we this power, however, there are other treasures far dearer to the heart which many would wish to evoke.

> "What hid'st thou in thy treasure-caves and cells,
> Thou hollow-sounding and mysterious Main?

---

daughters, and sold for charitable purposes. They have had the pleasure already of giving, in this way, about 110*l*. The books may be got at Messrs. Reeve and Benham's, Henrietta-street, Covent-garden; or at Mr. David Bryce's, Buchannan-street, Glasgow; or they can be sent by post, prepaid, from Saltcoats. The price, including postage, is 15*s*.

—Pale glistering pearls, and rainbow-coloured shells,
Bright things that gleam unreck'd of, and in vain.
  * * * * *
Yet more,—the billows, and the deeps have more!
High hearts and brave are gathered to thy breast!
  * * * * *
Give back the lost and lovely!—those for whom
The place was kept at board and hearth so long;—
The prayer went up through midnight's breathless gloom,
And the vain yearning woke 'midst festal song!
  * * * * *
To thee the love of woman has gone down;
Dark flow thy tides o'er manhood's noble head;
O'er youth's bright locks and beauty's flowery crown.
—Yet must thou hear a voice,—Restore the dead!
Earth shall reclaim her precious things from thee—
 Restore the dead, thou Sea!"—*Mrs. Hemans.*

Genus XXXVI. CALLITHAMNION, *Agardh.*

Gen. Char. Frond rosy or brownish-red, filamentous; stem either opake and cellular, or translucent and jointed, branches jointed, one-tubed, mostly pinnate (rarely dichotomous or irregular); dissepiments hyaline. Fruit of two kinds on distinct plants: 1, external tetraspores, scattered along the ultimate branchlets, or borne on little pedicels; 2, roundish or lobed, berry-like receptacles (favellæ) seated on the main branches, and containing numerous angular spores.—*Callithamnion* is from two Greek words signifying *beautiful little shrub.*—*Harvey.*

1. CALLITHAMNION PLUMULA, *Lyngbye*. Feathery Callithamnion.

Hab. In the sea, on every shore of Great Britain and Ireland, and yet far from being common on many shores.

It was first figured by Ellis more than eighty years ago, and then it does not seem to have been again observed till it was figured and described by Dillwyn, who got it at Swansea in 1802. His figure of it is good, but the colour is not sufficiently lively. It seems a deep-water plant, being got by dredging off the coast of Ayrshire and Arran. It is floated out in fruit in summer, and also so late as September. It is a captivating little Alga, the colour being a fine rosy-red, and the branches being beautifully pectinated, giving it a very feathery appearance. The capsules are small, but it is often dotted with large dark red favellæ.

2. CALLITHAMNION CRUCIATUM, *Agardh*.

Hab. Mud-covered rocks in the sea; rare. In England by Mrs. Griffiths, Mrs. Wyatt, Mr. Ralfs, Rev. Mr. Hore, Mrs. Gulson, Dr. Cocks; in Ireland by Dr. J. R. Harvey, Mr. W. Thompson, Mr. Andrews; and var. β by Professor Harvey at Miltown Malbay. Not got in Scotland.

It is a lovely plant, as may be seen by glancing at Plate CLXIV. in 'Phycologia Britannica.' It is easy to distinguish it from others by the tufts of branchlets at the top of each branch. By the aid of a lens the tetraspores at the base of the branches appear divided like a *cross*, whence the specific name.

3. CALLITHAMNION ARBUSCULA, *Lyngbye*.

Hab. On rocks and stones in the sea. Perennial. Common in many places. Rare on the east coast of Scotland, though found by Drs. Greville and Arnott, and by the Rev. Gilbert Laing. Not common on the west coast, though found in some abundance at Ballantrae. It seldom makes a good specimen for the herbarium, being in general too closely matted and too dark in colour.

When I had dismissed Miss *Dasya Arbuscula* with rather a frowning countenance, I was rebuked by looking at Dillwyn's figure, Pl. LXXXV. of 'British Confervæ.' It is lovely; and I am glad to make the *amende honorable*. He says, "Among the additions that have of late years been made to the list of British *Confervæ*, there is probably no species more beautiful or interesting than the present, which was discovered by Mr. Brown in the north of Ireland so long ago as 1800. The colour of this species when fresh appears to be a deep red brown; when dry it turns to a dull brown, tinged with green, wholly devoid of gloss." I must say that those few specimens I have gathered in a fresh state from the rock had not the lovely colour here spoken of; but I think I have seen specimens from other localities which were beautiful even when dried.

4. CALLITHAMNION BRODIÆI, *Harvey.*

Hab. On other Algæ; rare. Found at Forres by Mr. Brodie, of Brodie; Torquay by Mrs. Griffiths, Miss Cutler, and Mrs. Gulson; Cornwall by Mr. Ralfs; and by us at Saltcoats, though rarely.

Colour a brownish-red. The general outline of the frond is ovate, bearing a resemblance to the following.

5. CALLITHAMNION HOOKERI, *Agardh.*

Hab. On rocks and Algæ in the sea. Annual. Spring and summer. It is one of the most common on the Ayrshire coast, growing on coarse Algæ, and also on sand-covered rocks. It is found in great beauty on rocks in Arran.

It is from 1–3 inches high; colour brownish-red; branches spreading; a fine plant when the specimen is good.

6. CALLITHAMNION TETRICUM, *Agardh.* (Plate VII. fig. 25, a portion of the frond, natural size; and on the left, a plumule with a favella, and under it a plumule with tetraspores, both magnified.)

Hab. In the sea, on perpendicular faces of rocks, at half-tide level. Perennial. Common on the rocky coasts of England, and west and south of Ireland. I have not heard that it has ever been got in Scotland.

It is a large, rigid, shaggy plant, of a dull brownish-red colour. It is one of the coarsest of this family.

7. CALLITHAMNION ROSEUM, *Lyngbye*. (Plate VII. fig. 26, a branch, natural size; to the right a branchlet with tetraspores, magnified.)

Hab. On mud-covered rocks, and on Algæ. Annual. Summer. Yarmouth, Messrs. Turner and Borrer; Torquay, Mrs. Griffiths; Bantry Bay, Miss Hutchins; County Clare, Dr. Mackay.

This, I think, is rare in Scotland. I found a little scrap of it, which was named *Callithamnion roseum* for me by Sir Wm. J. Hooker, when I was beginning to collect Algæ, and I have got no more of it since. It is a beautiful plant.

8. CALLITHAMNION FLOCCOSUM, *Agardh*. (Plate VI. fig. 22, a portion of the frond, natural size; under it, on the right, a plumule with tetraspores, and on the left, a branchlet, both magnified.)

Hab. On submarine rocks, near low-water mark. Annual. Spring. Very rare.

This is a fine, red, loosely-branching plant, got only in the north of Scotland and in Norway. It was at first named in this country *C. Pollexfenii*, in honour of the Rev. Mr. Pollexfen, who found it in Orkney; but it was afterwards discovered that it had been named *C. floccosum* by Agardh. It has been gathered at Aberdeen by Dr. Dickie, to whose kindness I am indebted for some fine specimens.

9. CALLITHAMNION TURNERI, *Agardh*.

Hab. Parasitical on several marine Algæ.

Very large tufts of it are common in the island of Arran, on *Furcellaria fastigiata*. It was named *C. Turneri*, in honour of Mr. D. Turner, the author of 'Historia Fucorum,' by whom it was discovered. The fructification approaches to that of *Griffithsia*. *C. repens* is a variety of this plant.

10. CALLITHAMNION PLUMA, *Agardh*.

Hab. Generally on the stems of *Laminaria digitata*. Bantry Bay, Miss Hutchins; Appin, Captain Carmichael; Miltown Malbay, Professor Harvey.

A small plant, from a quarter to half an inch in height. A fine rose-red. It is rare. I have never seen it.

11. CALLITHAMNION BARBATUM, *J. Agardh*.

Hab. On mud-covered rocks. Perennial and very rare. A small tufted plant, found in Britain only by Mr. Ralfs and Rev. Mr. Berkeley.

12. CALLITHAMNION TETRAGONUM, *Agardh*.

Hab. On the larger Algæ. Annual. Summer. On the shores of England, Ireland, and also Scotland. It is occasionally found on the Ayrshire coast, and in the island of Arran.

Professor Harvey says, in his Phyc. Brit., Pl. CXXXVI., "When fully grown (four or five inches in height), this is one of the largest and most robust and shrubby British species of this charming genus, and seen under the water is an object of much beauty. In drying, though it sufficiently retains its form, it loses considerably in elegance, from the pressing together of the delicate quadrifarious ramuli, which become confounded with each other." It rapidly changes colour in fresh water, assuming a brilliant orange tint, and giving out a rose-coloured powder.

13. CALLITHAMNION BRACHIATUM, *Bonnem*.

I have been favoured with fine specimens of this by Miss A. Griffiths. It is very like *C. tetragonum*.

14. CALLITHAMNION BYSSOIDEUM, *Arnott*.

Hab. On other Algæ. Whitsand Bay, Dr. Walker Arnott; Devonshire, Mrs. Griffiths; Salcombe, Mrs. Wyatt; Strangford Lough, Mr. W. Thompson. We do not know that it has been found in Scotland. We have fine specimens of it from Mrs. Owens and Major Martin, got by them in Lough Swilly.

The stems are very slender, and the whole plant has a fine byssoid appearance. The colour is reddish-brown. This beautiful plant is intermediate betwixt *C. corymbosum* and *C. roseum*.

15. CALLITHAMNION POLYSPERMUM, *Agardh*.

Hab. Rocks in the sea. Annual. Spring and summer. In England by Mr. Griffiths, Mr. Borrer, Dr. W. Arnott; in Ireland by Miss Ball, Mr. R. Brown, Dr. Drummond, Mr. W. Thompson, Mr. Moore; Appin, Capt. Carmichael.

This is by far the most common *Callithamnion* in the west of Scotland. It is found on almost all the rocks within tide-mark. It is very abundant on the piers at Ardrossan, Saltcoats, Milport, and Largs. It is of a dull red colour; the tufts are globose; the filaments slender; the capsules very numerous, lining the inner faces of the pinnæ. The most interesting specimens I ever got of it were on the pier at Largo in Fife; they were not much above half an inch in height, but they were richly studded with the largest favellæ that even Mrs. Griffiths had ever seen on this species. Our Ayrshire specimens adhere very well to paper.

16. CALLITHAMNION TRIPINNATUM, *Agardh*.

Hab. On the perpendicular sides of the steep rocks at low-water mark.

It is extremely rare, having been found in Ireland only by Mr. M'Calla, at Roundstone Bay; and in England by Mr. Rohloff, at Plymouth. It has been found in France. It is as beautiful as it is rare. There is a lovely figure of it in Phyc. Brit., Plate LXXVII. It would not be very easy to distinguish it from one or two of the greatest beauties of this genus, were it not for a curious little pinnule at the axil of the pinnæ.

17. CALLITHAMNION GRACILLIMUM, *Agardh*. " Fern-leaved" *Callithamnion*, Mrs. Griffiths. (Plate VI. fig. 21, a tuft; on the left, a plumule, magnified; and under it a branchlet with a tetraspore, and another with a favella, both magnified.)

Hab. On mud-covered rocks. Mrs. Griffiths, pier at Torquay; Miss Warren, Falmouth; Mr. Ralfs, Milford Haven; Rev. W. S. Hore, Dr. Cocks, Mrs. Gulson, and Mrs. Spode, at Beaumaris.

Finely branched, with the tetraspores on the tops of the branchlets, and the favellæ at the base.

In looking at these beautiful works of God's hands, one would require an additional stock of epithets of admiration. Hear what Professor Harvey says respecting it:—"This extremely elegant plant, perhaps truly the *most graceful* of the very beautiful genus to which it belongs, was first gathered on the shores of France by M. Grateloup, who communicated specimens to the elder Agardh, by whom it was published in the year 1828. Shortly afterwards the indefatigable Mrs. Griffiths discovered magnificent specimens growing along the mud-covered base of the harbour pier at Torquay, in which locality it may be found in more or less plenty every summer. From Mrs. Griffiths it received the very appropriate name of '*Fern-leaf*,' aptly expressing the finely pinnated character of the branches, which do indeed closely resemble fairy ferns, so delicate, that it is altogether impossible in a figure to do justice to their beauty."

18. CALLITHAMNION THUYOIDEUM, *Agardh*.

Minute and beautiful and rare. Fine specimens have been found by Mrs. Gulson, in rock-pools in the Warren, opposite Exmouth; it has also been obtained in England by Mr. Borrer, Mr. Scance, Mrs. Griffiths, and Dr. Cocks; and in Ireland by Mr. W. Thompson and Mr. M'Calla.

19. CALLITHAMNION CORYMBOSUM, *Agardh*. (Plate VI. fig. 23, a branch, natural size; on the left a branchlet with tetraspores, magnified.)

Hab. On Algæ in the sea. Annual. June to September.

Not rare; from one to three inches high; frond with more or less principal stems, with long alternate branches, the branchlets of an obovate shape, though on the whole the plant is well marked by a level top or corymbose appearance of the branchlets. The capsules generally on the sides of the ramuli, the favellæ binate and large in the axils; the colour rose-red. It is a very handsome plant, adhering well to paper, and having, when dried, a glistering appearance. The finest specimens I have seen were got by Major Martin and myself, some of them on the wooden pier at Ardrossan, and others on numerous specimens of *Ascidia rustica*, which had attached themselves to the walls of the wet dock. It is found floating in the sea in Arran in September.

20. CALLITHAMNION SPONGIOSUM, *Harvey*.

Hab. On rocks and Algæ. Annual. Summer. In England, by Mrs. Griffiths, Miss Warren, Mrs. Wyatt, Rev. Mr. Horn, Mr. Rohloff, Mr. Ralfs, Mrs. Gulson, and Dr. Cocks; in Ireland, by Mr. Templeton, Dr. Drummond, Professor Harvey; in Scotland, by Major Martin, West Kilbride; at Saltcoats and Largo in Fife, by D. L.; by Miss White in Jersey.

This is dark-coloured and closely matted, and not very interesting in the state in which it is generally got by us in the west, but I have good specimens from Mrs. Griffiths.

29

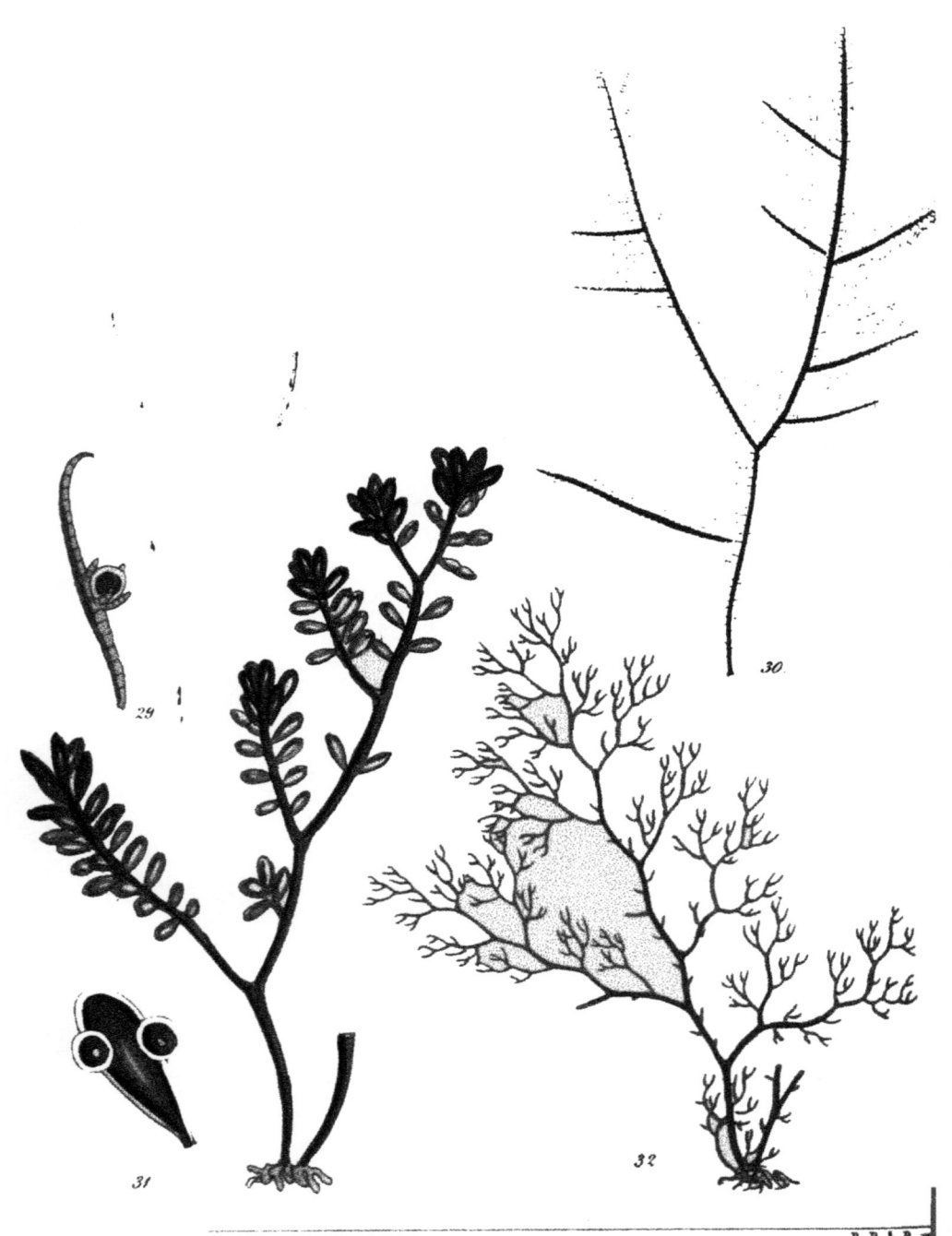

30

31

32

21. CALLITHAMNION PEDICELLATUM, *Agardh*.

Hab. On rocks in the sea. Not uncommon. Summer. It has been found in many places in England and Ireland. Iu Scotland it is rarer; it has however been found by the Rev. Mr. Pollexfen, in Orkney; by Major Martin and D. L. in Ayrshire; by D. L., jun., in November 1848, in fine fructification, in a rock-pool at Joppa, near Edinburgh. Very fine specimens were dredged by D. L. in Lamlash Bay.

22. CALLITHAMNION FLORIDULUM, *Agardh*.

Hab. On sand-covered rocks at all seasons. Abundant on the coast of Galway, where it was first observed by Dr. J. T. Mackay; by the Rev. Gilbert Laing, in the north of Ireland. We have not heard of its being found in Scotland, except by the Rev. Mr. Pollexfen, in Orkney.

The fructification was discovered by Mr. Ralfs on specimens found at the Land's End. It is very abundant in the west of Ireland, forming on rocks little cushions, which, when washed ashore in the end of summer, are called *figs* by the country people, and collected as manure.

23. CALLITHAMNION ROTHII, *Lyngbye*.

Hab. On rocks in the sea. Perennial. Winter. Not unlike the preceding, but smaller, and well distinguished from it by different fructification, as exhibited in 'Phycologia Britannica,' where both plants are figured in Pl. cxx. It is found both at Saltcoats and Ardrossan. At the former place the tufts are crowded, and nearly cover the rocks near high-water mark.

One is almost sorry, though he must own the justice and propriety of it, that in Phyc. Brit. *Callithamnion Rothii* is allowed to swallow up *Callithamnion purpureum* of Harvey's Manual. It had a pleasant classical interest attached to it. I have before me a little scrap of it which accompanied the following note from my kind friend Mr. Keddie of Glasgow. —" Last time I was in Iona, I made diligent search for *Conferva purpurea* of M'Culloch, the *Byssus purpurea* of Lightfoot, and the *Callithamnion purpureum* of Harvey; and was delighted to find it still growing where Lightfoot and M'Culloch saw it, near the Abbot Mackinnon's tomb. It occurs in purple patches, staining the lower part of the walls of the cathedral." It is also the *Conferva purpurea* of Dillwyn, whose figure is not so good as that in Phyc. Brit.

24. CALLITHAMNION SPARSUM, *Harvey*.

Hab. On old stems of *Laminaria saccharina* at Appin, Capt. Carmichael; at Miltown Malbay, on *Conferva rupestris*.

Even more minute than the preceding.

25. CALLITHAMNION DAVIESII, *Agardh*.

Hab. Parasitical on the smaller Algæ, such as *Ceramium rubrum*.

The forms elegant, pencilled tufts, which give a richly dressed appearance to the plant on which it fastens. Though it is a little fellow of only three lines in height, it is quite a gourmand, and has in Phyc. Brit. devoured, not only *Callithamnion lanuginosum* and my little favourite *Cal. secundatum*, but also *Callithamnion virgatulum*, though larger than itself. It is not uncommon on the west coast, especially at Portincross, in rock-pools on *Ceramium*.

Genus XXXVII. SEIROSPORA, *Harvey*.

Gen. Char. Frond rosy, filamentous; stem articulated, one-tubed, the articulations traversed by jointed filaments; branches jointed, one-tubed. Fruit, oval tetraspores disposed in terminal moniliform strings. Favellæ?—*Seirospora* is from two Greek words signifying *chain-seed.—Harvey*.

1. SEIROSPORA GRIFFITHSIANA, *Harvey*. (Plate VI. fig. 24, a fine branch, natural size; at the base, on the right, a branchlet with terminal tetraspores; and, on the left, the tetraspores removed, more or less magnified.)

Hab. On rocks and stones and shells, in deep water. Annual. Summer. Rare. Mrs. Griffiths, Torquay; Mrs. Wyatt, Salcombe; Rev. W. S. Hore, Plymouth; Mr. W. Thompson, Portaferry; Rev. J. H. Pollexfen, Orkney; D. L., in Lamlash Bay, Arran.

Seirospora Griffithsiana was formerly *Callithamnion seirospermum*, Griffiths. It was discovered by Mrs. Griffiths in the autumn of 1833. It has, however, been found to differ so much in fruit from *Callithamnion* as to lead Professor Harvey to form a new genus for its reception. In *Callithamnion* the tetrasporal fruit is borne laterally along the branchlets: in this the tops of the branchlets are converted into tetraspores. It is from one to three inches in height; colour a fine rosy-red; substance soft, gelatinous, adhering well to paper. It was found by us while dredging in Lamlash Loch, growing on living specimens of *Pecten opercularis.** It has been dredged by Major Martin on the

* Our boatmen were surprised at the avidity with which we grasped at whatever was growing upon the scallops. Among other things there were beautiful specimens of *Plumularia Catherina* and *Plumularia pinnata*, zoophytes which had certainly nothing in their appearance to recommend them in their state of collapse when removed from their native element. Taking one of the finest fronds of *P. pinnata*, I dipped it in water, and told them to look at it now. It had spread out into an elegant white plume; and re-

Ayrshire coast. A glance at our figure and at Phyc. Brit. Plate XXI., will lead you to say, "it is a lovely plant."

Genus XXXVIII. WRANGELIA, *J. Agardh*.

Gen. Char. Frond purplish or rosy-red, filamentous, jointed; filaments single-tubed. Fructification of two kinds: 1, tetra-

garding it with surprise, they said, they did not think there was anything so *bonnie* to be got in the bay. Another Arran fisherman, however, having brought up from the deep a fine specimen of *Plumularia myriophyllum*, Pheasant's-tail Coralline, took it home as a curiosity to his wife; and she being no less tasteful than her husband, planted it in earth in an old teapot, and, carefully watering it each day with fresh water, had the satisfaction of imagining that it grew a little under her fostering care. Be that as it may, it came unwithered into my possession, and its *vesicles* are embalmed in my friend Dr. Johnston's excellent 'History of British Zoophytes,' as being the first *vesicles* of this species that had ever been observed.

I am sorry that the scallops which yielded such rich crops of sea-weeds and zoophytes, have disappeared from the bay. The fishermen finding that they made excellent bait, had, in their greed, I suppose, exhausted the bed. There are different ways, however, of accounting for their disappearance. "Is Donald getting any clams (scallops) this year?" said I to a fisherman's wife. "Na, Na," answered she, "the clams and the fish have a' left our shore. Some bad men shot their nets on sabbath morning, and the fish and clams have a' forsaken the coast; and nae wonder," added she. "*Nae* wonder," responded I, "*nae* wonder!"—*Excursions to the Island of Arran, by D. L.* (Johnstone and Hunter, Paternoster-row, London.)

spores affixed to the inner sides of the ramuli (not confined to involucres); 2, gelatinous receptacles (favellæ) terminating the branches, surrounded by an involucre, and consisting of several clusters of pear-shaped spores, composted together.—The name in honour of Baron Von Wrangel, a Swedish naturalist. — *Harvey.*

1. WRANGELIA MULTIFIDA, *J. Ag.* (Formerly *Griffithsia multifida.*)

Hab. On perpendicular sides of deep marine pools. Frequent in the south of England and west of Ireland. Rare in Scotland. Caught (floating) by Mrs. R. M. Stark at Saltcoats. Found by Major Martin at Ardrossan, and dredged by him in Arran.

When this plant is seen in the water, it is remarkably beautiful, both in colour and structure. Its fine rose-red colour it soon loses in the open air or in fresh water; but the beauty of the structure still remains. The first specimen of it I ever saw, was given to me as a zoophyte. It is rare in Scotland. Once or twice during the summer it may be found by us, floating, but the specimens are always exceedingly fine, though not the fourth part of the size of Irish specimens, some of which would cover a quarto page.

" By the rushy fringed bank,
Where grows the willow and the osier dank,

> My sliding chariot stays ;
> Thick set with agate and the azure sheen
> Of Turkish blue, and emerald green
> That in the channel strays,
> Whilst from off the waters fleet,
> Thus I set my printless feet,
> O'er the cowslip's velvet head
> That bended not as I tread ;
> Gentle swain, at thy request
> I am here."—*Milton's Comus.*

Genus XXXIX. GRIFFITHSIA, *Agardh*.

Gen. Char. Frond rose-red, filamentous; filaments jointed throughout, mostly dichotomous; ramuli single-tubed; dissepiments hyaline. Fructification of two kinds, on distinct individuals: 1, tetraspores affixed to whorled involucral ramuli; 2, gelatinous receptacles (favellæ) surrounded by an involucre, and containing a mass of minute angular spores.—*Griffithsia*, so named by Agardh, in honour of Mrs. Griffiths of Torquay, the most distinguished of British Algologists.—*Harvey*.

1. GRIFFITHSIA EQUISETIFOLIA, *Agardh*.

Hab. On rocks at low-water mark. Perennial. Summer. Common in several parts of England, Ireland, and Wales; but very rare in Scotland. It was first figured more than seventy years ago by Lightfoot in his 'Flora Scotica,' from a specimen got by Mr. Yalden in the Frith of Forth; it is, therefore, not a little strange that it has not been found in any part of Scotland since.

2. GRIFFITHSIA SIMPLICIFILUM, *Agardh.*

This is like the preceding, but a more elegant plant. Its branches are attenuated to a sharp point, and it keeps its fine colour in drying. Coast of Norfolk, Rev. W. S. Hore, from whom I have a beautiful specimen.

3. GRIFFITHSIA DEVONIENSIS, *Harvey.*

While I refer for the figure of it to Phyc. Brit., Pl. XVI., I have great pleasure in quoting the following from that work:—

"This graceful little plant, which appears different from all the species of *Griffithsia* yet described, was discovered in the summer of 1840 by the Rev. Mr. Hore, at Plymouth; and in the autumn of the same year, added to the flora of Devonshire, by Mrs. Wyatt. I record this latter habitat in the specific name, because it affords me an opportunity, of which I avail myself, to connect the name of Mrs. Griffiths with that of the county whose shores she has so long and so successfully explored, where the best part of her life has been spent, and the natural history of which, in all its varied branches, her researches have so greatly advanced."

Ceramieæ.]   GRIFFITHSIA.   193

4.--GRIFFITHSIA BARBATA, *Agardh.*

Very rare. Found by Mr. Borrer on the beach at Brighton. Miss Turner has kindly sent me a specimen of it from Jersey.

5. GRIFFITHSIA CORALLINA, *Agardh.*

Hab. On rocks at low-water mark, or in deep pools. Annual. Summer. Very generally distributed.

This is one of the most attractive of our Algæ, and accordingly we find it noticed by our early botanists,—by Linnæus and Dillenius. Its fine red, glossy beads are decidedly coralline-like. I suspect that it is not common in Scotland; but when I say so respecting any plant, I wish it to be borne in mind that many of our richest localities have not been very scrutinizingly explored;—much, for instance, of the rich Carrick shores, in Ayrshire; much of Wigtonshire, about the Mull of Galloway, where my worthy friend the Rev. Mr. Lamb allows no rare *land*-plant to escape his notice, but who has not yet kept so keen a look-out for rare *sea*-plants; much also remains to be done about the Mull of Kintyre in Argyleshire, at Macrihanish Bay, and Dunaverty* Bay, perhaps the richest habitats in Scotland,

* *Dunaverty* means *the hill of slaughter*, and the mournful passage of history which gave rise to the name may be found in Sir W. Scott's 'Tales of a Grandfather.'

and where some rare plants have already been found by the Rev. Mr. Lambie.

It has been once got at Ardrossan. It has been obtained in Arran by Mrs. Balfour and Major Martin, as well as near Cromarty by Miss Allardyce. I have dredged it in Lamlash Bay, where it was found on the persecuted *Pecten opercularis*: quite willing was I to return the scallops to the sea after stripping them of their outer furniture. This, instead of being robbery, was mercy, for I am sure these sea-weeds and zoophytes must have been a sad incumbrance to them in those merry evolutions in which the scallops delight to indulge, when they skim through the water in mystic dance.*

6. GRIFFITHSIA SECUNDIFLORA, *J. Agardh*.

"This noble species, one of the finest of the section to which it belongs, was added to the British Flora by the Rev. W. S. Hore, who discovered it in August 1846, on rocks at extreme low-water mark at Bovisland, near Plymouth."—Phyc. Brit., Pl. CLXXXV. When the first edition

* The scallops dance most merrily. When I first saw some of them at their gambols in a tide-pool, I thought they were young fishes; but I found that they were the young of *Pecten opercularis*. By opening and suddenly shutting their valves, they skim rapidly along several yards, when they repeat the operation.

was published I was acquainted with this rare and beautiful plant only by the plate and description of it in 'Phycologia Britannica,' but I have since been favoured with a fine specimen of it by the Rev. W. S. Hore.

7. GRIFFITHSIA SETACEA, *Agardh*.

This is by far the most common of the *Griffithsiæ*. It is found in deep rock-pools. It is common in England and Ireland. It is not common in Ayrshire; but is found in Arran, and very abundantly in the south of Kintyre. It is very common in the Frith of Forth, from which specimens rich in fruit have been sent to me in July by D. L., jun. The first specimen I got with its involucres raised on club-shaped stalks was found by the Rev. Gilbert Laing, near Portobello. I scarcely knew what it was.

It is very rigid when fresh; but it soon loses this rigidity when exposed to the air, or when put in fresh water. In fresh water it gives all the symptoms of sudden and violent death. The membrane containing the fine carmine colouring matter bursts with a crackling noise. The plant yields its heart's blood, and dies; yet even in death it is beautiful. Professor Harvey says that it stands confinement well; that a tuft placed in a closed bottle of sea-water, at the end of more than two years' confinement, was as fresh and healthy as when taken from the sea. It stains paper of a fine carmine, which remains unaltered for many years.

Genus XL. SPYRIDIA, *Harvey*.

Gen. Char. Main filaments inarticulate, cartilaginous; beset with jointed ramuli; dissepiments opake. Fructification: 1, trisporous capsules with colourless margins, clustered round the bases of the ramuli; stalked gelatinous receptacles, with membranaceous pericarps, often surrounded by an involucre of short ramuli, containing two or three masses of roundish granules.—The name is from a Greek word, signifying a *basket*, in allusion to the appearance of the receptacles.—*Harvey*.

1. SPYRIDIA FILAMENTOSA, *Harvey*.

Hab. Submarine rocks, near low-water mark. Perennial. Summer. Southern coasts of England; rare. It has been found in Devonshire by Mrs. Griffiths, Miss Cutler, Mrs. Gulson, and others.

Till this was found lately by Mr. Ralfs in Wales, it was thought to be confined in Britain, to the south of England. It has very recently been found by Mrs. Spode at Beaumaris. It is found in Jersey by Miss Turner and Miss White.

Genus XLI. CERAMIUM, *Roth*.

Gen. Char. Filaments articulated, mostly dichotomous, reticulated with veins; dissepiments opake. Fructification double: 1, capsules, with a membranous pericarp, simple or lobed, generally subtended by one or two short ramuli, and containing numerous angular seeds; 2, oblong granules, partially imbedded in the joints of the lesser ramuli.—Name from a Greek word signifying a *little pitcher*, in allusion to the shape of the capsules in some of the species.—*Harvey*.

It is not long since our British list contained only two or three species of *Ceramia;* now we have, in the Phyc. Brit., fourteen enumerated.

1. CERAMIUM RUBRUM, *Agardh*. (Plate VIII. fig. 29, is a portion of *Ceramium rubrum,* natural size, and the figure to the left is a branchlet with a favella magnified.)

Hab. On rocks and Algæ. Perennial. Summer. Autumn.

This plant is very common everywhere. It is from six to twelve inches long; the capsules are globular, subtended by short ramuli; colour, from a fine red to a dirty white. So variable is it, that the young botanist thinks he has half-a-dozen species at least in his *vasculum,* when they all turn out to be this Protean *Ceramium rubrum*.

2. CERAMIUM DIAPHANUM, *Agardh*. (Plate VII. fig. 27,

is a very good figure of *Ceramium diaphanum,* natural size. On the right there is a magnified figure of a joint with whorled imbedded tetraspores. On the left, an involucrated favella, magnified.)

Hab. On rocks and sea-weeds. Winter and summer. Common.

This is a handsome plant, and much admired on account of the beauty and regularity of its jointings. It is a smaller and slenderer plant than the preceding; differing from it chiefly in having the joints colourless, and the dissepiments darkly coloured. The capsules are near the tips of the branches. It is a variable plant, like the former. Its most beautiful state is when the joints are a fine rose-red, and the dissepiments pure white.

3. CERAMIUM CILIATUM, *Ducluz.*

Hab. Rocks and corallines in the sea.

This may be distinguished from *Cer. diaphanum* by its greater rigidity; by the whorls of prickles with which the joints are furnished, and by the apices being very involute. The finest specimens I have are from Fingal's Cave, in the Island of Staffa. Very good specimens have been found on mud-covered rocks at Saltcoats, opposite the Free Church.

4. CERAMIUM ACANTHONOTUM, *Carm.* (Plate X. fig. 38,

a tuft of *C. acanthonotum,* natural size; and the figure on the left is a branch magnified to show the *spines.*)

Hab. On rocks at half-tide level. Annual.

This may be known from *C. ciliatum,* which it resembles, and with which it has in general been confounded, by its colour, which is darker; by its tufts, which are shorter and more compact; but more especially by the solitary three-jointed spine which arms the outer margin of every joint. The tetraspores encircle the joint, and the *favellæ* ensconce themselves within the incurved ramuli.

In Ayrshire this species is by far more common than *C. ciliatum,* for which it was long taken. At Saltcoats, it is found at half-tide level on rocks covered with young mussels, seeming to like to grow among them. At Portincross and in Arran, it is found in abundance, and the tufts are more than double the size of the tuft in the Plate. It does not adhere very well to paper, unless when it is young.

5. CERAMIUM ECHIONOTUM, *J. Agardh.*

Hab. It is not uncommon. It has been found by Mrs. Griffiths, Rev. Mr. Hore, Mr. Rohloff, Miss Ball, Professor Harvey, and Mr. Wm. Thompson.

At the time I discovered this species on the wall of the pier at Saltcoats, its name was not found in any of the lists of British Algæ. It has been since described and figured in 'Phycologia Britannica,' Pl. CXLI. It is distinguished by having the coloured dissepiments armed with numerous, slender, irregularly inserted, subulate, colourless, one-jointed prickles.

6. *C. flabelligerum*, J. Ag.   8. *C. strictum*, Grev. & Harv.
7. — *pellucidum*, Grev. & H.

9. CERAMIUM DECURRENS, *Kütz.*

In May 1846, when waiting on the quay at Largo, in Fife, for the steamer to convey me to Edinburgh, I employed myself in picking up Algæ. Among others, I found a very beautiful specimen of *C. decurrens*, which about the same time had been observed for the first time in this country by Mrs. Griffiths. I was at a loss whether to call it *C. rubrum* or *C. diaphanum*. When the figure of it lately appeared in 'Phycologia Britannica,' I was no longer at a loss. I have since found it in beauty at Ayr.

10. CERAMIUM GRACILLIMUM, *Grev. & Harv.*

This beautiful little plant, so far as I know, has not yet been observed in Scotland. I have specimens of it through the kindness of Miss Cutler and Dr. Cocks. It seems intermediate between *C. diaphanum* and *C. nodosum*.

11. CERAMIUM NODOSUM, *Grev. & Harv.*

Hab. Sandy shores, often at the roots of *Zostera marina*. Annual. Summer and autumn.

Though slender, it is rigid, and its forkings are very patent. The tetraspores are in a row, on the outside of the branchlets, and the *favellæ*, which are rare, at the tips of the branchlets. In England, by Mrs. Griffiths and Miss Kirkpatrick; in Ireland, by Miss Ball, Miss Gower, Mrs. Ovens, Professor Harvey, Mr. William Thompson, and Mr. M'Calla. It has been found occasionally floating in Saltcoats Bay, and has also been found on other Algæ on the pier-wall at Troon.

12. CERAMIUM FASTIGIATUM, *Harvey*.

This is a lovely species. The first specimen I ever saw of it was from Mrs. Griffiths, marked *very rare*. The next specimens were from Dr. Greville and D. L., jun., who got them at Joppa, near Edinburgh. It was first correctly distinguished by Mrs. Griffiths, who found it at Torquay. The filaments are dichotomous and level-topped; the colour in England is described as dark purple; in Scotland, the colour is rather light and lively purple. It is decidedly rare.

13. CERAMIUM DESLONGCHAMPSII, *Chauv*.

Hab. On rocks and Algæ between tide-marks. Annual.

Spring and summer. Pretty common in England, Scotland, and Ireland.

This species is quite common on the coast of Ayrshire. I long thought that it was a dark-coloured variety of *Cer. diaphanum*. It is distinguished from that species by its colour, the straight tips of the branches, and still more by the fruit, those clustered *favellæ* which burst from the stem and branches. It is found in rich tufts in early summer on the pier at Saltcoats. It has been gathered by D. L., jun., at Leith.

14. CERAMIUM BOTRYOCARPUM, *Greville*.

Hab. On rocks at low water. Annual?

This was discovered by Miss Amelia Griffiths at Torquay in 1844. It bears the same relation to *Ceramium rubrum* that *Ceramium Deslongchampsii* does to *Cer. diaphanum*. It is distinguished from *C. rubrum* by clusters of favellæ without involucres, bursting from the branches and branchlets, like clusters of grapes. It was found by D. L., jun., at Saltcoats, in September; and I found it at Ardrossan in November, rich in fruit, dark in colour, and retaining its colour in drying, and adhering well to paper. This makes me question its being annual. Like *C. rubrum*, it is a very variable plant, and some of the varieties are beautiful. The finest I have seen were found by Major Martin at Ardrossan, and by Mr. Keddie at Oban and Iona. It has been found by Mrs. Gatty at Filey; by Miss Allardyce at Cromarty; and by Mrs. Tyers at Bristol.

Genus XLII. MICROCLADIA, *Greville.*

Gen. Char. Frond filiform, compressed, distichously branched, traversed by a wide, articulated tube, surrounded by large, coloured, angular, radiating cells; external coat formed of minute reticulated cellules. Fructification of two kinds, on distinct individuals: 1, tetraspores immersed in the ramuli; 2, sessile, roundish receptacles (favellæ), having a pellucid limbus, containing numerous minute angular spores, and surrounded by several short simple involucral ramuli.—The name is from two Greek words, signifying a *small branch.*—*Harvey.*

1. MICROCLADIA GLANDULOSA, *Grev.* (Plate VIII., fig. 32, portion of the frond, natural size; figure on the left, branchlet with favella, magnified.)

Hab. On rocks, and on Algæ, at low water. Annual. Summer. Very rare. Mrs. Griffiths, Torquay; Miss Warren, Falmouth; Mr. Ralfs, Teignmouth; and specimens larger than usual, being $2\frac{1}{4}$ inches long, have lately been found by

Mrs. Gulson on the Warren, opposite Exmouth, growing luxuriantly on other Algæ. Obtained on the east coast of Ireland, but not in Scotland.

Microcladia bears some resemblance to *Ceramium rubrum*, but it has no external joints. The structure of it, as described in Phyc. Brit., Pl. xxix., is very curious.

Genus XLIII. PTILOTA, *Agardh*.

Gen. Char. Frond inarticulate, linear, compressed, or flat, distichous, pectinato-pinnate; the pinnules sometimes articulate. Fructification of two kinds, on distinct individuals: 1, tetraspores attached to, or immersed in, the ultimate pinnules; 2, roundish, clustered receptacles (favellæ) surrounded by an involucre of short ramuli.—*Ptilota*, from a Greek word, signifying *pinnated.*—*Harvey.*

1. PTILOTA PLUMOSA, *Ag.* (Plate VIII. fig. 30, a branch, and to the right, a pectinated pinnule, magnified.)

Hab. On the stems of *Laminaria digitata*. Perennial. Summer and autumn. Common.

This is a very handsome plant, and a general favourite. It is very common in Scotland. In England it is rare, and even unknown on the southern shores. It is a very variable plant. Some of our finest specimens were brought to us by

Rev. Gilbert Laing from Orkney. The largest specimen we ever saw was in Brodie's collection of Algæ, now the property of Professor Walker Arnott. It measures twenty inches in breadth by sixteen in height.

2. PTILOTA SERICEA, *Gmelin*. (Plate VII. fig. 28, natural size; to the left a magnified plumule.)

Hab. On the perpendicular sides of rocks, between tidemarks. Rarely on the stems of *Fucus serratus*. Perennial. Summer and autumn. Very common. Found also on the Atlantic shores of Europe and on the east coast of North America.

The difference betwixt this and *P. plumosa* may be seen by comparing the frond and pinnule in Plate VIII. with the frond and pinnule in Plate VII. It long ranked only as a variety of *P. plumosa*, and we are glad that it is restored to its original dignity of a species, the *Fucus sericeus* of Gmelin. We think it well entitled to the rank. *P. plumosa* is never found on rocks, always on *Laminaria digitata; P. sericea* is almost always found on rocks, and never that I know of on *L. digitata*. *Ptilota plumosa* is greatly infested by *Membranipora pilosa*: this zoophyte is scarcely ever seen on *P. sericea*, though in spring it is often almost covered by *Striatella arcuata*, a parásite which I have never observed on *P.*

plumosa. "The true difference, however, must be placed in the different structure of the ramuli, these being much more simple in the present plant." The specific name, *sericea,* expresses the soft and silky nature of the plant. The colour is naturally a dark blackish-brown. It is greatly improved in appearance when it assumes a pinky hue, and this may be helped on by exposure to the air. It is then a beauty.

Family VIII. GLOIOCLADIA.

"Thus Nature varies: man, and brutal beast,
And herbage gay, and scaly fishes mute,
And all the tribes of heaven, o'er many a sea.—
Search where thou wilt, each differs in his kind,
In form and figure differs."—*Lucretius.*

Genus XLIV. CROUANIA, *J. Agardh.*

Gen. Char. Frond gelatinous, filiform, consisting of a jointed, single-tubed filament, whose joints are clothed with dense whorls of minute multifid ramuli. Fructification of two kinds, on distinct individuals: 1, "favellidia, subsolitary near the apex of the ramuli, affixed to the base of the whorled ramelli, and covered by them, containing, within a hyaline membranaceous perispore, a subglobose mass of minute spores;" 2, obovate tetraspores of large size, affixed to the bases of the ramuli.—The name is in honour of the brothers Crouan, of Brest, celebrated among French Algologists.—*Harvey*.

1. CROUANIA ATTENUATA, *J. Agardh*.

Hab. Parasitical on the smaller Algæ. On *Cladostephus spongiosus* at Salcombe, Mrs. Wyatt; near Penzance, Mr. Ralfs. Annual. Summer. Very rare. See fine figures of it, natural size and magnified, in Phyc. Brit., Pl. CVI.

Genus XLV. DUDRESNAIA, *Bonnem*.

Gen. Char. Frond cylindrical, gelatinous, elastic; axis composed of lax net-work of anastomosing filaments, coated with a stratum of closely-combined longitudinal fibres; the periphery of horizontal, dichotomous, moniliform filaments. Fructification of two kinds, on different individuals: 1, globular masses of spores (favellidia) attached to the filaments of the periphery; 2, external tetraspores borne on the filaments of the periphery, generally terminating the ramuli.—The name is in honour of M. Dudresnay.—*Harvey*.

1. DUDRESNAIA COCCINEA, *Bonnem*.

Hab. Southern shores of England and Ireland. Very rare.

This is *Mesogloia coccinea* of Harvey's Manual. We have beautiful specimens of this rare plant from Rev. W. S. Hore and Dr. Cocks. The finest specimen of it we have seen was from Mrs. Gulson, Exmouth. Since the former edition was published, it has been found in Scotland by D. L., jun., in Arran, and about the same time by my daughter, Mrs. R. M. Stark, at Belhaven, near Dunbar.

2. DUDRESNAIA DIVARICATA, *J. Agardh.*

Hab. On stones and shells, not uncommon. Summer.

It is much branched; the colour pale red, or reddish brown; the substance tender and gelatinous. Mrs. Griffiths says, "The structure is very remarkable; the frond appears to be made up of tufts of fibres, radiating from a centre; each tuft, when separated in water under a glass, resembling a double aster, or sea-anemone. In the centre of the petal-like fibres are masses of purplish grains."

This species has been found in England and Wales by Mrs. Griffiths, Miss Warren, Mr. Ralfs; in Ireland by Miss Gower, Professor Harvey, Mr. W. Thompson, Mr. M'Calla; in Scotland by Mr. Brodie, Capt. Carmichael, Rev. Mr. Pollexfen; it has been found also at Saltcoats, and in the island of Arran. I have had specimens of it from France, from Jersey, and from Ireland, and I have seen several figures of it; but none that equalled in beauty the specimens that were got in Arran, which are firmer in substance,

33.

and of a richer brown colour, than any I have seen from other localities. It seems a northern plant. About the month of September it is got in beauty at Whiting Bay in Arran.

Genus XLVI. NEMALION, *J. Agardh.*

Gen. Char. Frond cylindrical, gelatinoso-cartilaginous, elastic, solid, axis columnar, dense, composed of closely-packed, longitudinal, interlaced filaments, whose alternate ramuli are moniliform and coloured. Fructification, globular masses of spores (favellidia) attached to the filaments of the periphery.—The name signifies a *crop of threads.*

1. NEMALION MULTIFIDUM, *J. Agardh.*

It somewhat resembles *Dumontia filiformis.* Though said to be not uncommon, we have never gathered it, and the only specimen we have is from the Rev. W. S. Hore.

2. NEMALION PURPUREUM, *Harv.* This was formerly *Mesogloia purpurea.* It has been found by Mrs. Griffiths and by Miss Cutler. Most beautiful specimens have lately been found by Mrs. Gulson at Exmouth, some of which she has sent to me, along with a drawing of the largest she ever found. It is magnificent—like a noble gnarled oak that had stood the wintry blasts of hundreds of years.

Genus XLVII. GLOIOSIPHONIA, *Carmichael*.

Gen. Char. Frond cylindrical, tubular, gelatinous; the periphery composed of a thin stratum of longitudinal, interlaced fibres, clothed externally with short, horizontal, branched, moniliform filaments. Fructification, spherical masses of spores (favellidia) immersed in the moniliform filaments to whose base they are attached.—The name signifies *a viscid tube.*—*Harvey*.

1. GLOIOSIPHONIA CAPILLARIS, *Carmichael*.

Hab. On submarine rocks and in tide-pools, near low-water mark. Annual. Summer. In England, by Hudson, Mrs. Griffiths, Miss Warren, Sir T. Frankland, Rev. H. Davies, Mr. Ralfs; in Ireland, Miss Hutchins, Dr. Drummond, Miss Gower, Mr. M'Calla; in Scotland, Captain Carmichael, D. L., and D. L. jun.; Jersey, Miss White and Miss Turner.

This is still a rare plant, even in England, and yet more so in Scotland, where it had been discovered only by Capt. Carmichael till it was found on the coast of Ayrshire. To see that it is a beautiful plant, we have only to look at the admirable drawing of it in Phyc. Brit., Pl. LVII., and to two excellent figures of it in Turner's 'Historia Fucorum,' all three true to nature in the different phases which it assumes. So rare was it forty years ago, when Mr. Turner published his standard work, that he wrote of it in the following manner:—

"The whole catalogue of British *Fuci* does not contain a single plant that has been so much involved in doubt as *F. capillaris*; for, being known only by Hudson's short and unsatisfactory description, and being a plant of unfrequent occurrence, even on the British shores, to which alone it appears to be confined, its very existence as a distinct species has been considered so questionable, that neither Dr. Goodenough and Mr. Woodward in their 'Observations on the British Fuci,' nor Mr. Stackhouse in his 'Nereis Britannica,' nor Professor Gmelin in his edition of the 'Systema Naturæ,' have ventured on introducing it; and even at the time of publishing the 'Synopsis of the British Fuci,' I had seen nothing more than two small specimens, so that I could say little respecting it which might be satisfactory either to my readers or myself. I have therefore felt a particular pleasure at being now enabled to remove it beyond the reach of doubt, through the kindness of my friends, Sir Thomas Frankland and Rev. H. Davies, both of whom, by communications with Hudson himself, know it to be the plant designed by that author, and both find it upon their own shores."

It has been found by D. L., jun., in rock-pools at Corrie and at Corriegills in Arran.* It has been found by myself on the Ayrshire coast at Saltcoats and Ardrossan. The first time I observed it was in Saltcoats Bay at low water, growing on shale. As I was in danger of being surrounded by the returning tide, I snatched in haste a small portion from a large patch, thinking it was some common thing with rather an uncommon aspect. On floating it in fresh water, spreading it on paper, and exposing it to the air, I was surprised to see it changing in a short time from a dull brownish-red to a fine dark crimson. One of my family, by wading into deep water, and catching the plants with his toes, got still finer specimens, which, being treated in the same manner, assumed even a richer hue. I soon found that it was this rare plant *Gloiosiphonia capillaris*. It has been carefully looked for every summer since, but it seems uncertain and capricious. Its season is limited, from the middle of June till the middle of July. Immersion in fresh water seems to bring sudden death on this as on some other sea-*plants*, but this change, as we see, only adds to its beauty.

* Vide 'Excursions to the Island of Arran,' by the Author, p. 274.

'Twas death,—and yet, than *life* more lovely !
The eye, turned heavenward, beamed with more than hope;
Yea, told of rapture; gleamed,—and closed in death.
On that calm brow, as Parian marble pale,
Rests truer dignity than ever clothed
The brow of potent king : peace, too, is there,
The perfect peace of God, for war is o'er,
Gained is the palm : the victor wears the crown !—*D. L.*

This rare and beautiful plant has lately been sent to us by Lady Keith Murray, who found it at Stonehaven on the east coast of Scotland.

Genus XLVIII. NACCARIA, *Endlicher.*

Gen. Char. Frond cylindrical or flat, filiform, solid; rose-red; central cellules large, empty, those of the surface minute. Ramuli composed of jointed, dichotomous, verticillate filaments. Fructification, groups of spores (favellidia) contained in swollen ramuli.—The name is in honour of F. L. Naccari, an Italian botanist, and author of 'Algologia Adriatica,' and other works. —*Harvey.*

1. NACCARIA WIGGHII, *Fries.*

Hab. On marine rocks at and beyond the limit of the tide. Annual. Summer. Very rare. It was discovered by Mr. Lilly Wigg (whence the specific name) on the Norfolk shore about the year 1790, and first described by

Mr. D. Turner. It has since been found by Mrs. Griffiths and Mr. Borrer, and recently by Mrs. Gulson in great abundance at Exmouth; by Miss Hutchins, Professor Harvey, and Mr. W. Thompson, in Ireland. It has been found for the first time in Scotland by D. L., jun., Macrihanish Bay, Argyleshire.*

Genus XLIX.   CRUORIA, *Fries.*

Gen. Char. Frond gelatinoso-coriaceous, forming a skin on the surface of rocks, composed of vertical, tufted, simple, jointed filaments, set in a gelatinous matrix; one of the joints of each filament greater than the rest. Fructification, tetraspores lying at the base of the filaments.—The name is from the Latin word for *blood* or *gore,* because the plant looks like a *blood-stain* on the rocks.—*Harvey.*

* "There are few naturalists, indeed," says Mr. Dawson Turner, "to whom the marine botany of England is under greater obligations than to my friend and original instructor in this department of science, Mr. Lilly Wigg. As a proof of which it will be sufficient to mention that six of the plants already mentioned in this work ['Historia Fucorum'] were first brought to light by his industry and acuteness of observation. It was he who first discovered the subject of the present plate, which I had a particular pleasure in describing under his name. It is an elegant species, both as to form and colour, and of remarkably unfrequent occurrence. For many years only three specimens, and those gathered at distant intervals, were known to exist."—Turner's 'Historia Fucorum,' page 84.

1.—CRUORIA PELLITA, *Fries.*

Hab. On smooth exposed rocks and stones between tide-marks. Fruiting in February. Perennial.

Though this plant has not been very generally observed, it is probable that it is widely distributed on our shores. It is very common on the west coast, and I must have seen it a hundred times, without ever thinking that it was a plant, till it was named by Professor Harvey among other things that I had sent for his inspection. My specimens were chiefly found on the roots of *Laminaria digitata*, which it often covers to some extent with a fine dark brownish-red skin, like japan. It takes its specific name, *pellita*, from its *skin-like* appearance. It is well worth while to consult Phyc. Brit., Pl. CXVII., where it is analysed, for it is surprising to see the curious filaments of which this *skin* is composed.

Family IX. NEMASTOMEÆ.

Genus L. IRIDÆA, *Bory.*

Gen. Char. Frond flat, expanded, carnose or gelatinoso-carnose, more or less of a purplish-red colour. Fructification, globules of roundish seeds, imbedded between the two coats of the frond.—*Greville.*

1. IRIDÆA EDULIS, *Bory.*

Hab. On marine rocks near low-water mark. Perennial. Fruiting in winter. Very common. When young it makes a pretty good specimen for the herbarium, adhering to paper. When old it becomes very dark in drying, and does not adhere. Stackhouse tells us that it is eaten in the south-west of England by the fishermen, after they have pinched it between hot irons, when it is said to taste like roasted oysters. If at all used in Scotland, it is after being roasted in the frying-pan. But though it should be loathed by men, to many of God's creatures it is very savoury; to crabs, mollusks, &c., it must be dainty food, for I have scarcely ever seen a full-grown specimen that was not perforated by these animals, like a flag that had long stood "the battle and the breeze."

Genus LI. CATENELLA, *Greville.*

Gen. Char. "Fronds filiform, somewhat compressed, creeping, throwing up numerous branches, contracted as if jointed in a moniliform manner, composed interiorly of branched filaments, radiating from the centre." — *Greville.* "Fructification: 1, spherical spores; 2, solitary, oblong tetraspores immersed in the periphery."—*Harvey.*—Generic name signifies a *chain,* in allusion to the *chain* or necklace form of the frond.

1. CATENELLA OPUNTIA, *Greville.* (Plate X. fig. 39 *a,* frond with fructification, magnified; *b,* plant, natural size.)

Hab. On rocks within high-water mark. Perennial. Not uncommon in England, Scotland, and Ireland. In Scotland we have gathered it from the rocks at Ardrossan in the west, and on the pier at Kessen ferry, in Ross-shire, near Inverness.

This little plant, seen on the rocks, is rather insignificant, like dwarf specimens of *Chylocladia articulata,* though more lurid in colour; neither does it make any great appearance in Stackhouse's plate. There is a good figure of it in Turner's 'Historia Fucorum,' a still better in our plate, with one of the kinds of fruit, and an excellent one in Pl. LXXXVIII. of Phyc. Brit., wherein both kinds of fruit, long unknown, are well represented. It has, however, to Algologists been a very troublesome little fellow, pushing its nose, like Paul Pry, into not a few of their genera. Driven from one, it took shelter in another, till Dr. Greville, in pity for its manifold sufferings under the alien act, gave it a permanent abode. He says, in his admirable 'Algæ

Britannicæ,' to which we have been so much indebted, and where there is a fine figure of it : " It has successively held the title of *Ulva, Fucus, Rivularia, Gigartina, Chondria, Halymenia, Lomentaria,* and, lastly, *mirabile dictu,* of *Chondria* in Sprengel's 'Systema Vegetalium.' I have endeavoured,—not, I think, without sufficient cause,—to afford this almost universal trespasser something more like a 'local habitation and a name.' " Its tetraspores were discovered by Professor J. Agardh, and both kinds of fruit have been detected by the keenly-scrutinizing eye of Mrs. Griffiths. It owes its specific name to its resemblance to the jointed leaves of *Cactus Opuntia.*

Family X. SPONGIOCARPEÆ.

"He looks abroad into the varied field
Of Nature, and, though poor, perhaps, compared
With those whose mansions glitter in his sight,
Calls the delightful scenery all his own.
His are the mountains, and the valleys his,
His the resplendent rivers."—*Cowper.*

Genus LII. POLYIDES, *Agardh.*

Gen. Char. Frond cartilaginous, filiform, cylindrical. Fructification, naked, spongy warts, of radiating filaments, among which are imbedded roundish clusters of wedge-shaped seeds, surrounded with a pellucid border.—*Greville*.

1. POLYIDES ROTUNDUS, *Greville*.

Hab. On rocks in the sea. Perennial. Autumn, winter, and spring. Rather rare. Found, however, in England, Scotland, and Ireland. In Scotland, by Dr. Greville, Frith of Forth; Dr. Richardson, near Dumfries; at Ardrossan, by Major Martin; by D. L., at Saltcoats in March, in fruit. It has lately been found abundantly in fruit by Lady Keith Murray at Stonehaven.

The root is an expanded disc, and by this, along with the fruit and rounded axils of the branches, it is distinguished from *Furcellaria fastigiata*. The colour is blackish-purple, becoming darker when dry. It does not adhere to paper.

Genus LIII. FURCELLARIA, *Lamour*.

Gen. Char. Frond cartilaginous, cylindrical, filiform, dichotomous. Fructification, terminal, elongated, pod-like receptacles, containing a stratum of dark oblong pear-shaped seeds, in the circumference.—The name signifies *a little fork.*—*Greville*.

1. FURCELLARIA FASTIGIATA, *Lamour*.

This is very common. It is often covered with large

patches of *Callithamnion Turneri,* and with *Tubulipora serpens.* When fresh it has a slight violet flavour.

Genus LIV. GYMNOGONGRUS, *Mart.*

Gen. Char. "Frond cylindrical, filiform, much branched. Fructification, naked warts composed of strings of cruciate tetraspores.—Name from the Greek, signifying *naked* and a *wart*, in allusion to the appearance of the fruit on the branches."— *Harvey.*

1. GYMNOGONGRUS PLICATUS, *Mart.*
Hab. On rocks in the sea. Perennial. Common.

Root, a small disc; fronds very numerous and matted together; they are horny, rigid, and thicker than a hog's bristles. In some localities they are ten inches in length; with us, from three to five; in drying, they do not adhere to paper. Very common on some parts of the coast of Ayrshire; much more common on the opposite islands of Cumbrae, where they are found on the shore in great reddish tufts.

2. GYMNOGONGRUS GRIFFITHSIÆ, *Mart.*
Hab. On rocks at low-water mark. Perennial. Winter and autumn. Rare. It is like a miniature specimen of

37.

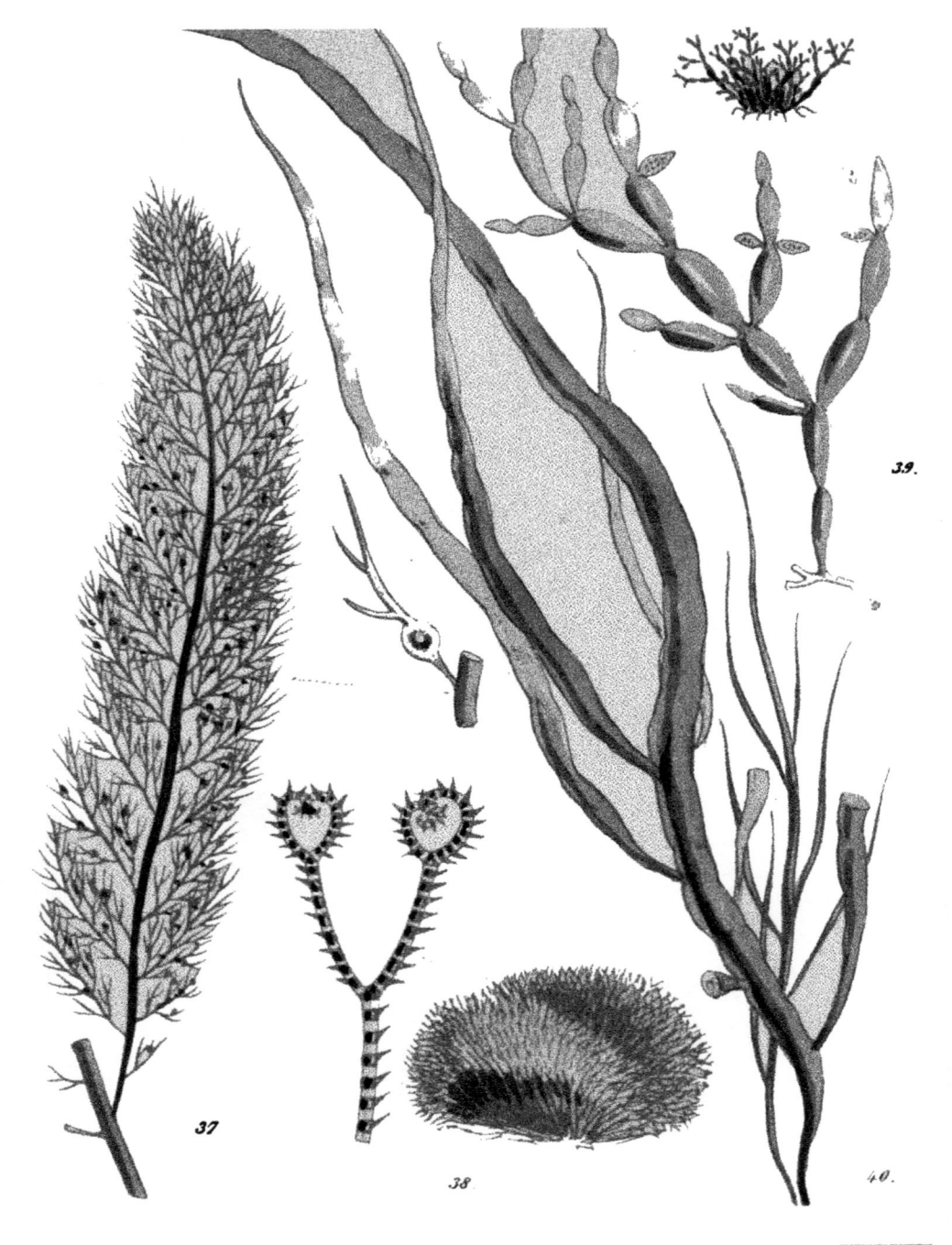

38.

39.

40.

Polyides rotundus, and was first correctly distinguished by Mr. Griffiths, whose name it bears. It is found by Mrs. Gulson at Exmouth.

Genus LV. CHONDRUS, *Stackhouse.*

Gen. Char. Frond cartilaginous, dilating upwards into a flat, nerveless, dichotomously divided frond, of a purplish or livid-red colour.—*Grev.* Fructification: 1, prominent tubercles (nemathecia) composed of radiating filaments, whose lower articulations are at length dissolved into spores; 2, tetraspores collected into sori immersed in the substance of the frond.—The name is from a Greek work signifying *cartilage.—Harvey.*

1. CHONDRUS CRISPUS, *Lyngbye.* (Plate IX. fig. 33, two varieties of this very variable plant, natural size; the larger one with *sori,* containing tetraspores.)

Hab. Rocks at low-water mark. Perennial. Spring. Very common and very variable. Turner figures ten, and Lamouroux thirty-five varieties.

At one time it was much in repute for furnishing by its gelatine a light easily-digested food for invalids, and, as the chief supply at first came from Carrageen in Ireland, it was called *Irish moss,* or *Carrageen.* The market price at one time was as high as 2*s.* 6*d.* per lb. Had it continued at that rate, it would have yielded more to the industrious inhabitants of the sea-shore than even a crop of their favourite potatoes. The fashion, however, has gone out, and the price has fallen; but the food prepared from it is as good as ever, and they who have tasted it once with good rich cream, will need no coaxing to partake of it a second time.

2. CHONDRUS NORVEGICUS, *Lamour*.

Hab. Rocky shores. Annual? September to November.

This is a pretty little plant; rather rare. Though called *Norvegicus*, it is oftener got in the south of England than in Norway. It is found in England, Scotland, and Ireland; occasionally only on the Ayrshire coast. The stem is cylindrical, the axils are patent, and the apices rounded. For an excellent figure of it, see Phyc. Brit., Pl. CLXXXVII.

Genus LVI. PHYLLOPHORA, *Greville*.

Gen. Char. Frond cartilaginous, or membranaceous, of a purple rose-red colour, plane, proliferous from the disc, furnished with a more or less imperfect or obscure midrib. Fructification: 1, capsules, containing a mass of minute roundish free seeds; 2, sori of simple granules in little foliaceous processes. (In two species nemathecia have been observed, but no granules.)—*Grev.*

1. PHYLLOPHORA RUBENS, *Greville*. (Plate IX. fig. 34, a good figure of *P. rubens* with fructification on the frond, natural size; and to the right at the base, a tubercle and a leafy process, with a *nemathecium*, magnified.)

Hab. In deep water. Perennial. Winter. It is found very generally on the British shores. It is not common on the coast of Ayrshire about Ardrossan, but is very abundant on the shores of the islands of Cumbrae, only a few miles distant. It has been found very large and fine in Arran by Mrs. Balfour.

Though it is not a favourite with me for the herbarium, because it is so rigid and rambling, and does not adhere to paper when dried, I allow no specimen to pass without strict examination, for it is a peculiar favourite of many species of zoophytes, and some of them of rare kinds; for instance, *Hippothoa catenularia, Hippothoa divaricata, Crisidia cornuta, Crisia chelata,* and rare *Lepraliæ,* &c. The old part of the frond is a dark brownish-purple; the young proliferous part of the frond is a lively rose-red.

2. PHYLLOPHORA BRODIÆI, *J. Agardh*.

Hab. Rocks in the sea; rare. Perennial? Winter and spring. Lossiemouth, Mr. Brodie; mouth of Bann, Mr. Moore; Bangor, Mr. W. Thompson; Devonshire, var.,

Mrs. Griffiths; Malahide, Mr. M'Calla. See 'Phycologia Britannica,' Plate xx.

I cannot refrain from quoting what is said by Mr. Dawson Turner respecting Mr. Brodie, in honour of whom this plant has its specific name. "The study of Natural History, independently of the advantages so nobly ascribed by Cicero to polite literature in general, that it nourishes our youth, delights our age, is an ornament in prosperity, and a comfort in adversity, may justly boast of a still superior object, in leading, and, indeed, forcing man to the admiration of the wisdom and the goodness of his Divine Creator in the contemplation of the works of his Almighty hand. In addition to this, it mixes itself also with the daily occurrences of social life, and gratifies the best feelings of our nature, by uniting in the bonds of friendship those whose pursuits were already the same; while, by permitting the names of its votaries to be affixed to plants, it records their zeal in its service, and touches one of the most powerful springs of human action. Among those who eminently deserve to be thus mentioned, stands forward the name of James Brodie, Esq., of Brodie, in Scotland, a man at once zealous in the pursuit, and liberal in the patronage, of universal science, and especially of the botany of Britain."

41.

42.

43.

44.

Fitch lith.

3. PHYLLOPHORA MEMBRANIFOLIUS, *Good. & Woodw.* On rocky shores. Common.

4. PHYLLOPHORA PALMETTOIDES, *J. Ag.* Rare. Coasts of Devonshire and Cornwall.

Genus LVII. PEYSSONELIA, *Crouan.*

Gen. Char. Frond membranaceous, orbicular or lobed, attached by the whole of its under surface.

1. PEYSSONELIA DUBYI, *Crouan.*

Hab. On old shells and stones in deep water. North of Ireland, Mr. W. Thompson; Birturbui Bay, Professor Harvey and Mr. M'Calla. Dredged by D. L. in Lamlash Bay, Arran.

After Professor Harvey had detected *Peyssonelia* in Ireland, he said, in a letter with which I was favoured, that I might be on the outlook for it when I went to dredge; and at the same time sent me specimens of it. On seeing them, it struck me that I had it on some old shells in my possession, and I sent him one which I had dredged in Lamlash Bay, with a dark brownish skin on it, which turned out to be true *Peyssonelia Dubyi*. It had been named specifically by M. Crouan, in honour of M. Duby. See Phyc. Brit.,

Pl. LXXI., which is much too light. It is not very rare on shells in deep water off the islands of Arran and Cumbraes.

Genus LVIII. HILDENBRANDTIA, *Menegh.*

1. HILDENBRANDTIA RUBRA, *Menegh.* On stones and pebbles, forming a thin membranous crust. See Phyc. Brit., Pl. XCVIII.

Family XI. GASTROCARPEÆ.

> " In Nature's all-instructive book,
> Where can the eye of reason look,
> And not some gainful lesson find,
> To guide and fortify the mind?"

Genus LIX. KALYMENIA, *J. Agardh.*

Gen. Char. Stem short, cylindrical, suddenly expanding into a roundish, subsimple, or irregularly cleft, somewhat lobed frond; favellidia densely scattered over the frond.

1. KALYMENIA RENIFORMIS, *J. Agardh.*

Hab. In deep shady pools at extreme low-water mark. Perennial? Summer and autumn. Discovered by Miss Everett in the Isle of Wight nearly half a century ago, and, when described by Mr. D. Turner in the 'Historia Fucorum,' it was regarded as exceedingly rare. Of late, however, it has been found in many places, especially in Ireland. Though found in Orkney by Rev. Mr. Pollexfen, and by D. L., jun., in Argyleshire, it seems very rare in Scotland. The splendid figure of it in Phyc. Brit., Pl. XIII., is thought rather dark in colour.

2. KALYMENIA DUBYI, *Harvey*.

This is a rare Alga, found by Miss Warren and by the Rev. Mr. Hore at Plymouth. It has not been met with in Scotland. In form it resembles *Iridæa edulis*, but in structure it is more like *Kalymenia reniformis*.

Genus LX. HALYMENIA, *Agardh*.

Gen. Char. Frond compressed or flat, pinky red, gelatinoso-membranaceous, consisting of a delicate membrane, whose walls are separated by a very lax net-work of jointed fibres; cells of the membrane minute, coloured. Fructification, masses of spores (favellidia) immersed in the frond, attached to the inner surface of the membranous periphery.—The name is from two Greek words, signifying the *sea* and a *membrane*.—*Harvey*.

1. HALYMENIA LIGULATA, *Agardh*. (Plate XIII. fig. 52, a good representation of two varieties of this plant.)

Hab. On rocks or stones near low-water mark, or more frequently dredged in deeper water. Annual. Summer. Frequent in the south of England, and not rare in the south of Ireland. Got in Orkney, and not very rare in the west of Scotland, where it has been repeatedly dredged off Arran and Cumbraes and the coast of Ayrshire by Major Martin and D. L. It has been found at Southend, Kintyre, by the Rev. Mr. Lambie. It is much more common in England and in Jersey.

When the colour is rose-red, as it often is, this is a beautiful plant, especially if the frond is broad; but it is very variable in form. At times the frond is quite narrow and dichotomous. For the fruit and structure, see Phyc. Brit., Pl. cxii.

Genus LXI. GINNANIA, *Mont.*

Gen. Char. Frond terete, dichotomous, membranaceo-gelatinous, traversed by a fibrous axis, from which slender, dichotomous, horizontal filaments radiate towards the membranous periphery; surface cellules hexagonal. Fructification, spherical masses immersed in the frond, affixed to the inner coating, composed of radiating filaments, whose apical joints are converted into spores.—Name in honour of Count Ginnani, an Italian botanist.—*Harvey.*

1.-GINNANIA FURCELLATA, *Mart.*

Hab. On rocks and shells in deep water. Annual. Summer. Rather rare.

This fine plant may seem to resemble the dichotomous varieties of *Halymenia ligulata,* but the structure and frond are different. Professor Harvey mentions a remarkable feature of its structure that has been little attended to, viz., an *axis* or internal rib somewhat like the midrib of a *Delesseria.* This plant has been got in England and Ireland; but it had not been got in Scotland till the autumn of 1850, when it was dredged in Arran by Mrs. Balfour, of Edinburgh, and gathered by her also at very low water on the rocks betwixt Lamlash and Corriegills. Major Martin likewise dredged it in Lamlash Bay.

Genus LXII. DUMONTIA, *Lamour.*

Gen. Char. Frond cylindrical, simple, or branched, membranaceous, tubular, gelatinous within, of a red or purplish-red colour. Fructification, globules of seeds attached to the inner surface of the membrane of the frond.—The name is in honour of M. Dumont, a French naturalist.—*Greville.*

1. DUMONTIA FILIFORMIS, *Greville.* (Plate X. fig. 40.)

Hab. Stones and rocks in the sea. Annual. Summer.

Common. On the coast of Ayrshire the twisted variety is the most common. I remember that it was one of the first I requested Sir W. J. Hooker to name for me, mentioning its pungent taste and smell as characteristic. See fine representations of the fructification in Dr. Greville's 'Algæ Britannicæ,' Plate XVII., and in Professor Harvey's 'Phycologia Britannica,' Plate LIX.

Family XII. COCCOCARPEÆ.

"He who, through nature's various walks, surveys
The good and fair, her faultless line portrays;
Whose mind, profaned by no unhallowed guest,
Culls from the crowd the purest and the best;
May range, at will, bright fancy's golden clime,
Or, musing, mount where science sits sublime,
Or wake the spirit of departed time."
Pleasures of Memory.

Genus LXIII. GIGARTINA, *Lamour.*

Gen. Char. Frond cartilaginous, either filiform, compressed, or flat, irregularly divided, purplish-red; axis or central substance composed of branching and anastomosing longitudinal filaments; the periphery of dichotomous filaments laxly set in pellucid jelly, their apices moniliform, strongly united together. Fructification double, on distinct plants: 1, external tubercles, containing on a central placenta dense clusters of spores (favellidia) held together by a net-work of fibres; 2, tetraspores scattered among the filaments of the periphery, or aggregated in dense, immersed sori.—The name is from the Greek word for a *grape-stone*, which the tubercles resemble.—*Harvey*.

1. GIGARTINA PISTILLATA, *Lamour.*

Hab. On rocks near low-water mark. Perennial. Winter. Very rare. Discovered by the Hon. Dr. Wenman in 1800. Found by Stackhouse, Brodie, Mrs. Griffiths, Miss Hill, Dr. M'Culloch, Dr. Jacob, Mr. Gilbert Sanders; by Miss Turner in Jersey. It does not seem to have been procured in either Scotland or Ireland.

It is a remarkable plant; the tubercles are large, generally near the point of the branch, which projects like a horn; the colour is a dull purplish or brownish red; it does not adhere to paper. A person who has seen the very correct figure of it in Phyc. Brit., Pl. ccxxxii., could not fail to know it, were he to light on so great a prize. I am glad that I have even better than the figure, having received specimens of this very rare plant from Professor Harvey, Rev. W. S. Hore, and Dr. Cocks.

2. GIGARTINA ACICULARIS, *Lamour.* (Plate XI. fig. 42, plant in fruit.)

Hab. Submarine rocks near low-water mark. Annual. Winter. Rare.

This bears some resemblance to the preceding; but it is a less robust plant. The tubercles are rare and smaller, and the dull purple colour becomes pink in fresh water. It is found in England by Mr. Rashleigh, Mrs. Griffiths, Miss Cutler, and Mrs. Gulson; in Ireland by Mr. W. Thompson and Professor Harvey.

3. GIGARTINA TEEDII, *Lamour*.

Hab. On rocks in the sea; very rare. Perennial. Ellery Cove and Tor Abbey rocks, by Mrs. Griffiths, to whom, as also to the Rev. W. S. Hore, I am indebted for a fine specimen of this rare Alga.

4. GIGARTINA MAMILLOSA, *J. Agardh*.

Hab. Rocks in the sea, near low-water mark. Perennial. Autumn and winter.

This species is as common as the others are rare; the fronds are from 3–6 inches long, cartilaginous, channelled; capsules on little stalks scattered profusely over the disc of the frond. We have found it very large in Arran, rolled up in the form of a ball. The finest specimens we have ever got were at Gourock, near Greenock. Till we met with it figured and new-named in 'Phycologia Britannica,' we regarded it as a *Chondrus*. It is gathered and employed for culinary purposes along with *Chondrus crispus*, or *Carrageen*. Indeed, the lady who first collected *Carrageen* in Ayrshire, to make blanc-mange, gave the preference to this species, saying, that it was more gelatinous than the other. I have seen the fronds in the Clyde more than an inch broad. The Plate cxcix. in Phyc. Brit. would be very good if it were darker.

Genus LXIV. GELIDIUM, *Lamour.*

Gen. Char. Frond, between cartilaginous and corneous, planocompressed, distichously branched, branches pinnate or bipinnate, pinnæ spreading or horizontal, obtuse, capsules spherical, immersed in the extremities of the ramuli.—*Greville.*

1. GELIDIUM CORNEUM, *Lamour.*

Hab. On rocks in the sea and in rock-pools. Perennial. Summer. On most of the rocky shores. It is found in great beauty by Miss Cutler at Budleigh.

"The varieties of this species," says Dr. Greville, "are almost endless, and some of them so singular, that without practical knowledge to guide us in our investigation they may be taken for very distinct species." I can vouch for the truth of this, for in a very beautiful variety sent to me by the Rev. Mr. Smith, of Borgue, I thought I had got the southern *Grateloupia filicina*. For good figures and descriptions, see 'Algæ Britannicæ,' Pl. xv., and 'Phycologia Britannica,' Pl. LIII. Dr. Greville describes about fourteen varieties. The prettiest variety I have ever seen has been sent to me by Mrs. Tyers, near Bristol. The branches are capillary, and the ramuli are "in the form of inversely lanceolate or ovate leaves, much attenuated at their insertion." It is the var. *clavatum* of Dr. Greville.

* 2. GELIDIUM CARTILAGINEUM, *Greville*.

Hab. On rocks in the sea. Perennial.

A magnificent plant, but a very doubtful native. The frond is from 12-18 inches long; the colour is a dark purple, but when it begins to be decomposed it is shaded with fine tints of scarlet, orange, yellow, and green. I remember seeing a good specimen framed as a painting, and exposed for sale in the Cowgate of Edinburgh. I procured two specimens of it from Dublin, where, *on dit*, it is occasionally found in the bay. I learn from Professor Harvey that it has lately been cast ashore at Ryde. Through the kindness of Professor Walker Arnott, I have a few specimens of it from Brodie's Alg. Herb.

Genus LXV. GRATELOUPIA, *Agardh*.

Gen. Char. Frond flat, more or less pinnate, membranaceous, flexible, solid, composed of densely interwoven, anastomosing, branching filaments, those of the periphery moniliform, short, and very strongly compacted together. Fructification: 1, globular masses of spores (favellidia) immersed beneath the peripheric stratum, communicating with the surface by a pore; 2, cruciate tetraspores, vertically placed among the filaments of the periphery, in sub-defined sori.—In honour of Dr. Grateloup, a French Algologist.—*Harvey*.

1. GRATELOUPIA FILICINA, *Agardh*.

Hab. On rocks and stones in the sea. Very rare. Discovered by Miss Cutler. Found by Mrs. Griffiths at Barrowcane; and by Mr. Ralfs, Mount's Bay, Cornwall; plentiful. The finest specimens I have ever seen were gathered by Mrs. Gulson at Exmouth. It has not been found in Scotland or Ireland. "The only British plant with which this is likely to be confounded is *Gelidium corneum,* to some varieties of which, especially *G. flexuosum,* it bears a very strong resemblance. Its softer and more membranous substance will generally distinguish it to the feel; and the microscope will point out a difference of structure."—*Harvey*.

For good figures, see 'Algæ Britannicæ,' Pl. xvi., and 'Phycologia Britannica,' Pl. c.

Family XIII. SPHÆROCOCCOIDEÆ.

> "Oh! what an endlesse work have I in hand,
> To count the sea's abundant progeny!
> Whose fruitful seede farre passeth those in land,
> And also those which wonne in the azure sky:
> And much more eath to tell the starres on by,
> Albe they endlesse seeme in estimation,
> Than to recount the sea's posterity,
> So fertile be the flouds in generation,
> So huge their numbers and so numberlesse their nation."
> <div align="right">SPENSER'S <i>Faery Queene.</i></div>

Genus LXVI. HYPNEA, *Lamour.*

Gen. Char. Frond filiform, cartilaginous, continuous, much branched, cellular; with a dense, more or less evident fibro-cellular axis, surrounded by several rows of angular cells, the innermost of which are largest, the outer gradually smaller to the circumference. Fructification of two kinds, on distinct individuals: 1, spherical tubercles (coccidia) sessile, or immersed in the ramuli, containing a mass of small spores on a central placenta; 2, transversely parted tetraspores imbedded in the cells of the surface.—*Hypnea*, an alteration of *Hypnum*, the name of a genus of mosses, in allusion to the mossy characters of some of the original species.—*Harvey.*

1. HYPNEA PURPURASCENS, *Harvey.* (Plate X. fig. 37, *a,* branch with tubercles, of the natural size; *b,* magnified portion of a branchlet with a tubercle or coccidium.)

For the reasons which have induced Professor Harvey to transfer this species to *Hypnea* of Lamouroux, see Phyc. Brit., Pl. CXVI.

Hab. In the sea on rocks and the larger Algæ. Annual. July to November. Frequent on the shores of England, Scotland, and Ireland. It is got at an early period of the season, and being then without tubercles, it is apt to be mistaken for some other plant. In this growing state the branches occasionally terminate in capillary tendrils, which twine around other Algæ. Lightfoot says, "The fructification appears like little grains or tubercles of a round or oval figure, hardly so big as the smallest pin's head." With us the oval tetraspores are as big as middle-sized pins' heads. When full grown, this plant is very bushy, and at times two feet in length; the stem as thick as a crow's quill, thickest at the middle, and attenuating towards each end; substance cartilaginous, but soft, and adhering to paper; colour brownish or purplish pink, becoming darker in drying. I preserve a very dark specimen as a memorial of a pleasant half-hour's algologizing in the sweet bay of Rosemarkiee, in Ross-shire.

Genus LXVII. GRACILARIA, *Greville.*

Gen. Char. Frond filiform, or rarely flat, carnoso-cartilaginous, continuous, cellular; the central cells very large, empty, or full of granular matter; those of the surface minute, forming densely packed, vertical filaments. Fructification of two kinds, on distinct individuals: 1, convex tubercles (coccidia) having a thick pericarp, composed of radiating filaments, containing a mass of minute spores on a central placenta; 2, tetraspores imbedded in the cells of the surface.—The name is from the Latin word signifying *slender.—Harvey.*

1. GRACILARIA ERECTA, *Greville.*

Hab. On sand-covered rocks, near low-water mark. Perennial. Fruiting in winter. Sidmouth and Torquay, Mrs. Griffiths; in Ireland, by Mr. W. Thompson, Mr. D. Moore, and Mr. M'Calla; in Orkney, by Rev. T. H. Pollexfen, Lieut. Thomas, and Dr. M'Bain.

This curious little plant is one of the numerous discoveries of Mrs. Griffiths. When in fructification it cannot be mistaken for anything else, as the densely-clustered tubercles on the branches, and the lanceolate terminal pods containing tetraspores, are sufficient to distinguish it. When not in fruit, it is like *G. confervoides* in a young state. As the Scotch specimens were not in fruit, some doubt hangs over them. In Plate XIV. of Alg. Brit. there are excellent figures of it, both of the natural size and magnified, with dissections of the fruit, &c.; and the same may be said of Plate CLXXVII. of 'Phycologia Britannica.'

2. GRACILARIA CONFERVOIDES, *Greville*. (Plate XI. fig. 44, branch of the natural size; conjoined with it there is a small portion of a branch with tubercles, magnified.)

Hab. In the sea on rocks. Perennial. In fruit from August to October. It is said to be of rather frequent occurrence on the British shores, and yet I do not know that it has been gathered in Scotland, except by the Rev. J. Macvicar, many years ago in the Frith of Tay, and by Brodie of Brodie, at Jossiemouth, in 1806, and by us in August 1850, dredged near Milport in the island of Cumbrae. It is more than a foot and a half in length, and easily known by its tubercles. Colour deep red. It is cartilaginous, and does not adhere to paper in drying.

3. GRACILARIA COMPRESSA, *Greville*.

Hab. Cast ashore from deep water, attached to coral, &c.

Very rare. Annual. Summer. Discovered by Mrs. Griffiths at Sidmouth. Found also by Miss Cutler in the south of England, and lately by Mrs. Gulsou, at Exmouth; and by Miss Turner, Jersey.

Substance when fresh, very tender, and succulent and brittle, becoming tough when dried; colour, dull red, becoming brighter when steeped in fresh water. In form, it very much resembles *Gracilaria lichenoides*, bleached specimens of which I have from my friend Mr. Gourlie. Having lost their colour, they make no show in the herbarium, but they appear very well on the table in the shape of *blanc-mange*, which is of a brownish-red colour. Mrs. Griffiths got some of our native *G. compressa* prepared for the table, and it answered as well as the foreign one; but it is too rare to be so employed, except by way of experiment.

Since writing the above, I resolved, with the remainder of Mr. Gourlie's Agal-Agal, or *G. lichenoides*, to try if it would, by putting it on gauze, form a lantern *à la mode de Chine*. The experiment was quite successful. I have also within these few days seen an edible swallow's nest, which had every appearance of being formed of mashed Agal-Agal. The form is oval, joined to the rock at one of the ends, truncated for that purpose, and very like a recep-

45.

46.

47.
a b

48.

B.H.&R.imp.

tacle for holding a watch. Some feathers of the birds were adhering to it still.

4. GRACILARIA MULTIPARTITA, *J. Agardh.*

Hab. On rocks and stones in muddy places in the sea, chiefly in estuaries, near low-water mark. Annual. August and September. Found by Miss Hill; Rev. W. S. Hore; Dr. Cocks, Plymouth; and by Mrs. Wyatt, Salcombe Bay.

This is a showy, but rare plant, being as yet found in Britain only on the southern shores of England. It is, however, widely distributed over the world. Its oldest name is *Fucus multipartitus* of Clemente. For a fine figure and description see Phyc. Brit., Pl. xv.

Genus LXVIII. SPHÆROCOCCUS, *Agardh.*

Gen. Char. Frond cartilaginous, much branched in a distichous manner, compressed and two-edged below, nearly flat upwards, the branches acute at the apex; capsules spherical, mucronate on little stalks fringing the smaller branches.—*Greville.*

1. SPHÆROCOCCUS CORONOPIFOLIUS, *Agardh.* (Plate XII. fig. 48, part of the frond, of the natural size, and a portion of a branchlet with tubercles, magnified.)

Hab. On rocky shores. Perennial. Summer and autumn. Common on the southern shores of England; not uncommon in Ireland. Rare in Scotland, where, however, it has been found by Dr. Greville in the island of Bute; by Major Martin at Ardrossan; by D. L. at West Kilbride, Ayrshire, and in the island of Arran.

From the similarity in colour, and from some resemblance in the outline, it may at times, in Scotland, have been passed over as *Plocamium coccineum*. It is, however, a much larger plant, being a foot and a half in length. It is one of the most beautiful of the British Algæ. When procured fresh, the colour is a fine rose-red. For excellent figures, with instructive dissections, see Pl. xv. of 'Algæ Britannicæ' of Dr. Greville, and Pl. LXI. of Professor Harvey's 'Phycologia Britannica.'

Genus LXIX. RHODYMENIA, *Greville*.

Gen. Char. Frond plane, membranaceous, fine pink or red, quite veinless, sessile, or with a short stem, which expands immediately into the frond. Fructification: 1, hemispherical scattered capsules; 2, minute, ternate granules, spreading over the whole, or some part of the frond (not in defined spots).—Name from two Greek words signifying *red membrane*.—*Greville*.

1. RHODYMENIA BIFIDA, *Greville.*

Hab. On rocks in the sea, and on Algæ. Summer. Annual. Frequent on the southern shores of England. In Ireland, where it is not rare, fine broad-fronded specimens have been got by Miss Hyndman and Mr. D. Moore. It has been got by Dr. Greville in Bute. Not uncommon on the coast of Ayrshire, where it has been often found by Major Martin and by D. L. It is most abundant in early autumn, when it is cast out in the form of round tufted balls, frequently the abode of the pretty zoophyte, *Valkeria cuscuta.*

A curious variety of this plant was gathered by my kind friend Mr. Keddie, respecting which Professor Harvey said that it was so different from the usual appearance, that had it come from a distant country he should have been disposed to exalt it into a species, under the name of *R. fimbriata.* It is, however, a very variable plant everywhere. It is as thin as a *Nitophyllum* in fruit, but it differs from *Nitophyllum* and even from other *Rhodymeniæ.*

2. RHODYMENIA LACINIATA, *Greville.* (Plate IX. fig. 36, plant of the natural size, with a small portion of the margin with imbedded tubercles or *coccidia,* magnified.)

Hab. On rocks, and on *Laminariæ*; generally in deep

water. Biennial. January to July. Frequent in Scotland, England, and Ireland. It is not common, however, on the coast of Ayrshire about Saltcoats and Ardrossan. It is frequent and very fine at Southend, Kintyre, at Dunaverty and Macrihanish Bay. The most beautiful specimens I have seen of it were gathered in the island of Mull by her Grace the Duchess of Argyll and Lady Emma Campbell, and not thought less lovely because prepared by persons whose great worth and numerous accomplishments give additional dignity to their high rank and station.

Its substance, being thick, is very different from *R. bifida*. When got among rejectamenta, and partly decomposed, the specimen is spotted with white, which has not a bad effect. See a fine figure of it, natural size, and magnified figures of both kinds of fruit, in Pl. cxxi. of Phyc. Brit.

3. RHODYMENIA PALMETTA, *Greville*.

Hab. On rocks, and on *Laminariæ*. Annual. Summer and autumn. Not uncommon on the British shores. Rare in the west of Scotland. Found by Mrs. Gibb at Saltcoats, and by D. L. near the lighthouse, Pladda, off Arran.

Substance rigid, and imperfectly adhering to paper; colour a fine pinky-red, retained in drying. For excellent descriptions and figures, see Alg. Brit., Pl. xii., and Phyc. Brit., Pl. cxxxiv.

4. RHODYMENIA CRISTATA, *Greville.*

Hab. Parasitical on the stems of *Laminaria digitata.* A northern species. Very rare. Annual. July. Found by Sir W. J. Hooker and Mr. Borrer at Wick in Caithness; in the Frith of Forth, Dr. Greville; at Berwick, Dr. Johnston. I have lately been much gratified by receiving from Professor W. H. Harvey, American specimens of this rare British plant. It varies a good deal in form. Some varieties of it are like small specimens of *Sphærococcus coronopifolius*.

5. RHODYMENIA CILIATA, *Greville.*

Hab. On rocks in deep water. Annual. Winter. Frequent on the southern shores of England, and in some places in Ireland. Rare in Scotland, where it was found by Lightfoot in Iona; and by Lieut. Thomas and Dr. M'Bain, in Orkney.

Substance thicker than any other *Rhodymenia*; colour, a deep red. I have a fine specimen from Miss White, Jersey. It has lately been dredged off Cumbraes by Major Martin, and off Arran by D. L.

6. RHODYMENIA JUBATA, *Greville.* (Plate XI. fig. 43, frond, natural size, and a magnified figure of a cilium with a tubercle.)

Hab. On rocky or gravelly shores. Annual. Summer. Frequent in the south of England, and in some places in Ireland. Said to be common in Scotland, but we have not found it so. Some dwarf specimens of it were found by Mr. R. M. Stark at Ballantrae, and it has been dredged off Arran by Major Martin and D. L.

8. RHODYMENIA PALMATA, *Greville*.

Hab. On rocks, and other Algæ. Very common. Annual or biennial. Winter and spring.

Instead of giving any further description of this plant, it is sufficient to say that it is *Dulse*, and every child who has been brought up on the sea-shore, is able to point it out to the new-fledged Algologist. There is no sea-weed more generally regarded as an article of food than Dulse. By the Highlanders it is called *Duillisg*, which, we learn on high authority, is a word compounded of two Gaelic words, *duille*, a *leaf*, and *uisgé*, *water*, i. e., the *leaf of the water*. From *uisgé* is derived the word *whisky*; and with the addition of *baugh*, *life*, we have the *usquebaugh* of the Irish (*aqua vitæ*), *the water of life*: with how much more propriety might it be called *the water of death*!

In some parts of Ireland the *Dulse* is called *dillisk*, which means still *the leaf of the water*, for *esk* means water:

hence we have so many rivers in Scotland named *Esk*, such as North Esk and South Esk, i. e., North-Water, South-Water. The Highlanders and Irish, as we have already stated, were much in the habit, before tobacco became so rife, of washing Dulse in fresh water, drying it in the sun, rolling it up, and then chewing it as they now do tobacco. How much better had it been for them had they stuck to the use of the less nauseous, less filthy, less hurtful Dulse. Indeed, instead of being hurtful, it is thought wholesome and not unpleasant, especially when it is eaten fresh from the sea, as is the case in the Lowlands. Dr. Greville mentions that it is the true *Saccharine Fucus* of the Icelanders. According to Lightfoot, it is used medicinally in the isle of Skye, to promote perspiration in fevers. In the islands of the Archipelago, it is a favourite ingredient in *ragouts*, to which it imparts a red colour, besides rendering them of a thicker and richer consistence. The dried frond, like many other Algæ when infused in water, exhales an odour resembling that of violets; and Dr. Patrick Neill mentions that it communicates that flavour to vegetables with which it is mixed.

Rhodymenia sobolifera was long ranked as a distinct species; but Professor Harvey, in 'Phycologia Britannica,' mentions it only as a variety, and I know that Dr. Greville regards it in the same light. It has been dredged by Mrs. Balfour in Arran. We have got it growing on *Fucus serratus.* It is cast out by the tide in great abundance on the shore of Holy Isle, Arran, and occasionally on the Ayrshire coast.

Genus LXX.   STENOGRAMME, *Harv.*

Gen. Char. Frond rose-red, leaf-like, nerveless, laciniate, cellular; the central cells large, transparent, in several rows, those next the surface minute, coloured, closely packed. Fructification linear, convex, longitudinal. The name is from two Greek words, signifying a *narrow line,*—alluding to the linear fructification.—*Harvey.*

1. STENOGRAMME INTERRUPTA, *Agardh.*

Dashed up from deep water. Annual. November. Very rare. Bovesand and Plymouth, Dr. John Cocks. Minehead, Somerset, Miss Gifford. Mount Edgecombe, Rev. W. S. Hore, to whose kindness I am indebted for fine specimens of it. Colour a fine, clear, pinky red, resembling much *Rhodymenia Palmetta* in external appearance.

Family XIV. DELESSERIEÆ.

"There's beauty in the deep:—
The wave is bluer than the sky;
And though the light shine bright on high,
More softly do the sea-gems glow,
That sparkle in the depths below;
The rainbow's tints are only made
When on the waters they are laid;
And Sun and Moon most sweetly shine
Upon the Ocean's level brine:—
There's beauty in the deep."

J. G. C. Brainard.

Genus LXXI. PLOCAMIUM, *Lamouroux*.

Gen. Char. Frond filiform, compressed, between membranaceous and cartilaginous, fine pink-red, much branched, branches distichous (alternately secund and pectinated). Fructification of two kinds: spherical sessile capsules, and lateral minute processes, containing oblong granules, transversely divided into several parts by pellucid lines.—The name is from a Greek word signifying *braided hair.*—*Greville.*

1. PLOCAMIUM COCCINEUM, *Lyngbye*. (Plate XII. fig. 46, portion of the frond, natural size; and at the base, on the left, a branchlet with a tubercle, magnified.)*

* "The description," says Mr. Turner, "given of this plant in the 'Flore Française,' is so characteristic that I am tempted to transcribe a part of it.

" ' Sa tige est très-rameuse, et toujours dans le même plan: l'ordre des

Hab. Common almost everywhere in the sea. Perennial. Summer and autumn. Though generally common, it is rather of unfrequent occurrence on the sea-shore about Ardrossan and Saltcoats, but very frequent at Ballantrae in Ayrshire. So abundant is it on the coast of Kintyre, Argyleshire, that it might be got in cart-loads; the same is the case on the coast of East Lothian, where it really is carted away for manure.

It has generally a peculiar appearance, according to the place where it grows. At Leith, where it is very frequent, it is more cartilaginous, more erect, and of a darker hue than the Ayrshire specimens; and the pattern, we would say, less *genteel*.* The Irish specimens are large and strong, as well represented in Phyc. Brit., Pl. XLIV.

ramifications est très-remarquable ; chaque rameau est légèrement flexueux, et n'émet de ramifications que du côté convexe : la première est un filet simple et pointu ; la deuxième est un filet qui a trois dents du côté antérieur ; la troisième est un filet qui a deux dents, et qui au lieu de la troisième dent pousse un filet muni d'une dent en dehors ; la quatrième est un filet qui n'a qu'une dent, la deuxième dent est devenue un filet à une dent, et la troisième un filet rameux. Après ces quatre ramifications il y a un espace vide, et la tige émet des rameaux semblables du côté opposé.' "

* Since writing the above, I have received from Mrs. R. M. Stark, Edinburgh, specimens gathered by her at Leith, as beautiful in colour and as *genteel* in form as any found in the west.

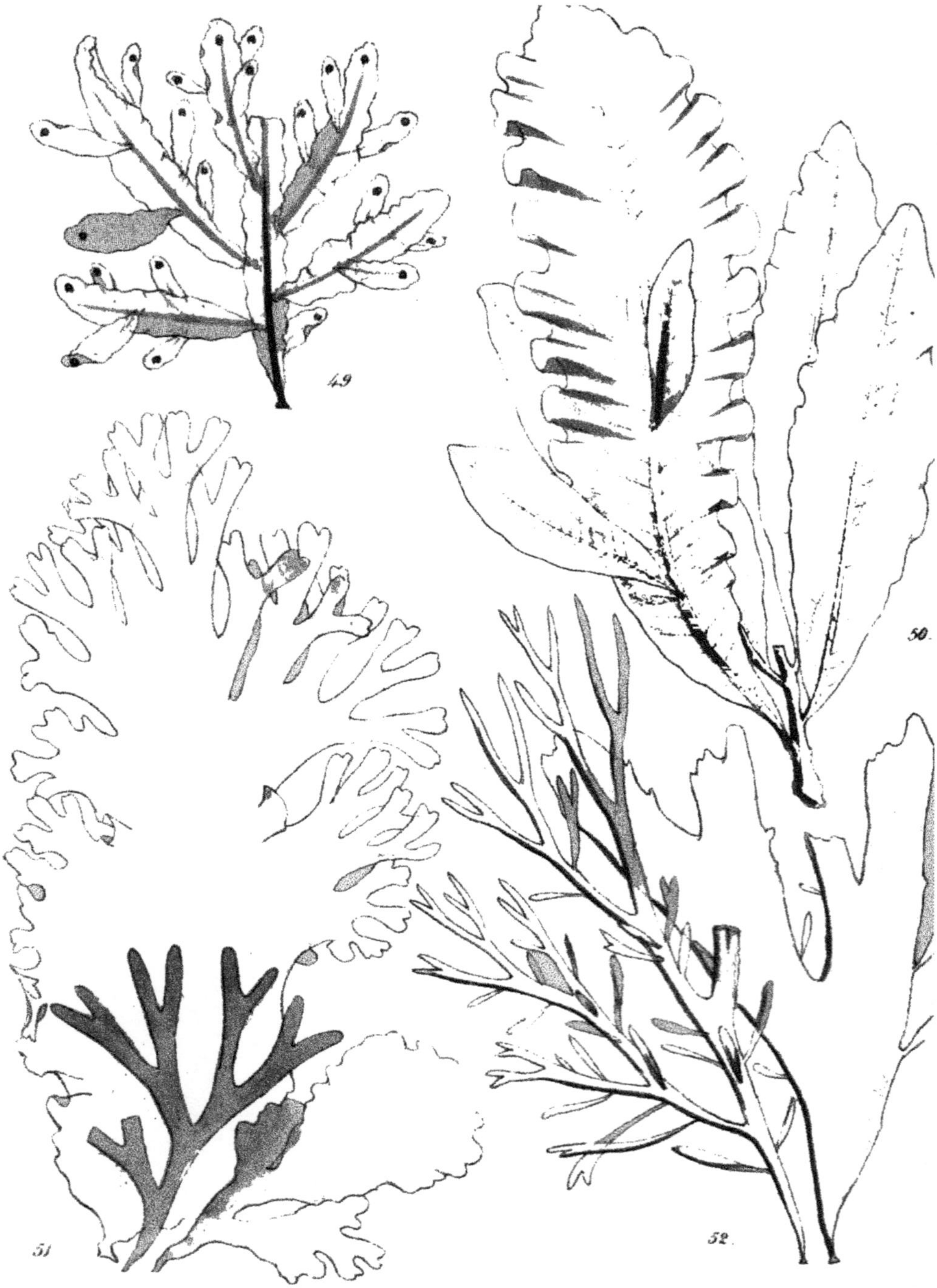
49
50.
51
52.

It is not easy to account for the variety of form which the plant assumes in different localities. Looking at a pretty specimen got from Miss White, in Jersey, through the kindness of Mr. Smith of Jordan-hill, we were going to ascribe its greater softness of substance and more flowing ramification, to the more genial climate; but turning our attention to a rich specimen from North Ronaldshay, we saw that the Orcadian was as flowing in ramification, and as flaccid in substance, as the native of the Channel Islands. A specimen gathered by Mr. Keddie at Iona, and another found by the Lady Emma Campbell in Islay, resemble those found in the north of Ireland. The one from Lady Emma Campbell acquires additional value from having a very rare zoophyte nestling among its lower branches,— *Alecto granulata* of Milne Edwards. Dr. Johnston describes and figures it in his most interesting 'History of British Zoophytes.' It had previously been procured only by Mr. Couch, in Cornwall, and by Mr. William Thompson in Ireland. Keep a good look-out, my young friends, for precious parasites on what comes from deep water, or new localities.

The colour of *Plocamium coccineum* is beautiful, especially when exposed a little to the sun after a shower of rain; it then becomes a lovely crimson. "One of the most charming and symmetrical Algæ in the world, extremely common, and a universal favourite," says Dr. Greville. "A well-known, abundant, and beautiful species, and an especial favourite with amateur weed-collectors, and manufacturers of sea-weed pictures," says Professor Harvey.

Genus LXXII. DELESSERIA, *Lamouroux*.

Gen. Char. Frond rosy-red, flat, membranaceous, with a percurrent midrib. Fructification of two kinds: 1, capsules containing a globular mass of seeds; 2, ternate granules forming definite sori, in the frond, or in distinct foliaceous leaflets.—The name is in honour of a noble French patron of science, the Baron B. Delessert.—*Greville.*

1. DELESSERIA SANGUINEA, *Lamour.* (Plate XIII. fig. 50, plant with leaflets from the midrib.)

Hab. On rocks and also on other Algæ. We have found it on the roots of *Laminaria digitata.* It is biennial.

The fructification, of two kinds, may be seen on the naked midrib in December and January. The stem is about an inch in length, considerably thicker than a crow-quill. It is occasionally divided into more than a dozen fronds, from two to four inches broad, and from six to ten in length;

generally acute at the tip, but at times rounded. The margin is more or less waved.

A beautiful variety, mentioned by Professor Harvey as sent to him by D. L., was found floating at Saltcoats by Miss M'Leish; she got it only once, but it was a great bunch, of a dozen branches, some of the fronds being eight inches in length, and five and a half in breadth. The peculiarity of this remarkable variety was, that it was lobed somewhat like *Delesseria sinuosa,* with a midrib in each lobe. Another specimen of the same kind was got by Miss Ramsay, of Glasgow, at Gourock. It was a splendid specimen, of a very rich colour, with three large lobed fronds, the largest being nine inches and a half by five.

There is in the possession of my friend Major Martin of Ardrossan, a magnificent frond of *D. sanguinea,* which Sir William J. Hooker said was one of the largest he had ever seen. The single frond, or leaf, is thirteen inches long and eight inches broad! This gentleman's collection of Algæ is most splendid, and he has almost as much pleasure in displaying these rich prizes of peace, as in showing his well-earned trophies of war. He lately received one of the Peninsular medals as an acknowledgment of his eminent services in the Peninsula, during a number of years; he received also a gold medal for having commanded the 45th Regiment at the battle of Toulouse; he was most dangerously wounded at Ciudad Rodrigo, having been shot through the body while fighting at the top of the main breach.

Stackhouse speaks of *D. sanguinea* as not being common, and as seldom found entire. On the coast of Ayrshire in spring and early summer, it may be got on the shore in abundance, and quite entire, after a stiff breeze. In January it might puzzle the young Algologist, as nothing then remains but the red stem, and midrib beset with fruit of both kinds, on separate plants. In February it makes an interesting specimen with young leaves, more than an inch in length, springing from the midrib, mixed with the fruit. In March the fruit has disappeared, but the leaves are then three or four inches in length, and almost an inch in breadth, of a fine fresh glossy pink hue, and of a lighter and more delicate tint than when they are full-grown. When full-sized, the fronds are generally, on the Ayrshire coast, about seven inches long, and an inch and a half in breadth, though often much larger, and the colour is then a splendid rich pinky red. In its spring and summer dress it is a lovely plant, and it is not wonderful that it should be a universal favourite. Mr. Dawson Turner says: " In the elegance of its appearance, and the exquisite colour of its most delicately-veined leaves, this beautiful *Fucus* so much excels all its congeners, that it carries away the palm with no less justice from the vegetables of the ocean, than the rose, the flower of the poets, from its rivals in the garden." In Phyc. Brit., Pl. CLI., Professor Harvey says: "This fine plant, whether we regard the splendour of its colour or the elegance of its form, is entitled to high rank in the Oceanic Flora, and notwithstanding its common occurrence on all our shores, is never seen without attracting admiration. In favourable localities it reaches to a very large size,—and such specimens are among the most beautiful vegetable objects in nature." The substance of the leaves is delicately membranous. They are often plaited along the margin, and this gives additional beauty to dried specimens by varying the hues of the fine crimson pink. We may mention that the fronds, if allowed to remain long in fresh water, give out much of their colouring matter, and are consequently paler when dried, and have less also of that glossy shining aspect which otherwise characterizes them.

2. DELESSERIA SINUOSA, *Lamour*.

Hab. On the larger Algæ. Common.

It is questioned whether it is biennial. I have observed

it during all the winter on the stems of *Laminaria digitata*, though it is then in a ragged state, and does not adhere to paper in drying. Mr. Turner, in his 'Historia Fucorum,' Pl. xxxv., shows it when advanced to the second year of growth. Stackhouse, also, in 'Nereis Britannica,' has a figure of it; but neither of them gives us the plant in its most beautiful state, as found in Ireland and the west of Scotland. I have before me an Irish specimen, the frond of which is six inches in length, and as many in breadth, and I have seen them larger. I have also before me a Scottish specimen gathered at Gourock, by my friend Miss Ramsay, of Glasgow, which, though not the largest, is the richest and most beautiful one I ever beheld. It consists of three fronds of large size, shaped like an oak-leaf, the colour of which is a fine dark brownish-purple. It is rather remarkable that this, and the very splendid *Delesseria sanguinea*, which I have already mentioned, should have been found so far up, where the fresh water of the Clyde must have a considerable effect. Perhaps they were driven from their moorings in the Holy Loch, which is nearly opposite, and in the shelter of which they might expand themselves; or, as Miss Ramsay thinks, from the rocky point at Portkill in the parish of Roseneath.

This is a very variable plant in its appearance, according to its age; the leaves, when very young, being oval. The substance is thin and membranaceous, adhering tolerably to paper in drying, unless when old; the colour when fresh is red, but not so fine a red as the preceding; it becomes much darker in drying, unless decomposition has begun, in which case there is a fine variety of tints, red and green, and yellow and white.

3. DELESSERIA ALATA, *Lamouroux*.

Hab. In the sea, chiefly on *Laminaria digitata*.

It is questioned whether it is perennial. I can answer for it that it is at least biennial. It is got all winter in a ragged, sapless state, when it does not adhere to paper in drying. Early in March it begins to grow, and in a short time the lower part of the frond is dark and rigid, and the upper is light-coloured, and fresh and limber. In the end of March 1849, I got specimens in this state, some of them with capsular fruit on the young part of the frond, and others with ternate granules thickly imbedded in little leafy processes at the very tips of the frond, having a rich appearance.

This is by far the most abundant *Delesseria* on our western shores. Even in its most common state it is a handsome plant. It is occasionally found on the coast of Ayrshire,—and much more frequently on the Irish coast,—in such a state as to be truly magnificent. In some specimens collected by Dr. Drummond, Belfast, the frond or winged membrane is half an inch broad. I have a specimen of this description gathered by Mrs. Lyon at Glenarm, and more than one procured by Mrs. Ovens and Major Martin in Lough Swilly, of which the winged membranes are not only broad, but the whole plant is larger than usual, and the colour a fine rich dark purple, so that at first a person would take it for a member of a nobler tribe than our every-day acquaintance, *Delesseria alata*.

4. DELESSERIA ANGUSTISSIMA, *Griff.*

Hab. On the stems of *Laminaria digitata*. Perennial. Winter and spring. Rather rare. Lossiemouth, Mr. Brodie; Aberdeen, Dr. Dickie; Cornwall, Mr. Ralfs; Orkney, Rev. Gilbert Laing and Rev. J. H. Pollexfen; island of Islay, Lady Emma Campbell.

This is one of the *Algæ vexatæ*. When found by Mr. Brodie in the north of Scotland about forty years ago, and sent to Mr. Turner, they agreed in considering it a variety of *D. alata*, naming it *D. angustissima*. In deference to a person of remarkably sound judgment, and who is seldom mistaken, Professor Harvey described it in his Manual as *Gelidium? rostratum*; in 'Phycologia Britannica,' he figures and describes it as *Delesseria angustissima*; in his excellent Plate LXXXIII. the figures of the fruit exactly correspond with those of *D. alata,* except that the tubercles or capsules in his figure are in sharp-pointed, axillary ramuli: whereas in my specimens of *D. alata* they are in the rounded or dichotomous tips of the frond. I have beside me a specimen from Dr. Dickie, Aberdeen; another from the Rev. Gilbert Laing, Orkney; and a third gathered by Lady Emma Campbell, in the island of Islay: the latter is a fine rambling plant, in which the compressed edge of the frond begins to be membranous.

5. DELESSERIA HYPOGLOSSUM, *Lamouroux.*

Hab. On rocks and on other Algæ. Annual. Summer. Frequent on the shores of England and Ireland, and not rare in Scotland.

The distinguishing characteristic of this pretty species is the repeatedly proliferous leaflets from the midrib. It is very variable. The frond is sometimes half an inch in breadth, but this is only in Ireland, where it reaches its maximum of beauty. In Scotland it is not uncommon. In some seasons it is abundant on the coast of Ayrshire, and the size is often about equal to the beautiful figure, Pl. II. in 'Phycologia Britannica.' But at times in Ayrshire and in the island of Cumbrae the frond is remarkably narrow, not above half a line in breadth. I have received fine specimens from the Rev. Gilbert Laing from Orkney, and I have one magnificent specimen procured by Major Martin in Lough Swilly, of which the frond is so large that it might be mistaken for a young specimen of *Delesseria sanguinea*.

6. DELESSERIA RUSCIFOLIA, *Lamour.* (Plate XIII. fig. 49, plant with tubercles on the leaves and leaflets, natural size.)

Hab. On rocks, and sometimes on the stems of *Laminaria digitata*, and on other Algæ. Annual. Summer and autumn. It is rare in Scotland. It has twice been procured, floating, in fine condition, by Miss M'Leish, in the sea at Saltcoats. Miss White, Isle of Portland. By far the finest specimens we have ever seen were from the Rev. W. S. Hore and Dr. Cocks. They were of large size, and the colour was even richer than that of *D. sanguinea*, which is saying a great deal.

"How calm, how beautiful comes on
The stilly hour when storms are gone;
When morning winds have died away,
And clouds beneath the glancing ray

53.

55.

b.

a.

Delesseriea.]

> Melt off, and leave the land and sea
> Sleeping in bright tranquillity,—
> Fresh as if day again were born
> Upon the rosy lap of morn!"

Genus LXXIII. NITOPHYLLUM, *Greville.*

Gen. Char. Frond plane, delicately membranaceous, rose-coloured, reticulated, wholly without veins, or very slight vague ones towards the base. Fructification, hemispherical capsules imbedded in the substance of the frond, and ternate granules forming distinct scattered spots.—*Greville.*

1. NITOPHYLLUM PUNCTATUM, *Greville.* (Plate XIII. fig. 51, plant with spots of granules scattered over the frond.)

Hab. In the sea, attached to various Algæ. Annual. Summer. It seems to be found on most of the shores of England and Ireland, and on the Scottish shores as far north as Orkney.

On the coast of Ayrshire, we would say, that in general it is rather rare; yet there are seasons when it is very abundant. The summer of 1847 was one of these. In the summer of 1848 it was very rarely observed. In the summer of 1850, again, it was often met with. It has the property, like *Nitophyllum versicolor*, of assuming an orange hue in fresh water, and of recovering its original colour when dried. Fresh water acts also on it, as on *Griffithsia setacea*. My daughter, Mrs. Stark, when floating a newly-collected specimen in fresh water, cried out, " Hear how it *fizzes.*" In the struggles of death it made this hissing, crackling sound, but it was soon over. The specimens got in the west of Scotland, though very beautiful, are seldom more than five inches in length by four in breadth. With what surprise should we gaze were a gigantic Irish specimen to be floated over to us, such as those found by Mr. D. Moore at Cushendall Bay, in the west of Ireland—five feet long, by three feet wide! We have heard of mermaids reclining on a rock, combing their beautiful flowing locks. Were they ever, in the pride of their hearts, to think of assuming female attire, what robe could be more appropriate and becoming than a spotted *Nitophyllum*, with its finely-lobed margin encompassing their neck, and turned back, *à la Vandyke*, on their shoulders? Where is the merman who would not be fascinated with their graceful appearance in this new costume? The colour of this marine mantle is a beautiful pale rose-pink. Its beauty is much increased by the darker-hued capsules and sori with which it is spotted.

It now includes *Nitophyllum ocellatum.* Dr. Greville's Pl. XII. in Alg. Brit., with fructification, is very instructive. Professor Harvey devotes Plates CII. and CIII., Phyc. Brit., to the illustration of the fruit, textures, and varieties of this interesting and beautiful Alga.

2. NITOPHYLLUM LACERATUM, *Greville.*

Hab. In the sea, on rocks and Algæ; to which it can attach itself by its edges, and by little roots from the under side of the frond, so that it can scarcely be torn off without being injured.

It is a very valuable plant; the colour is much darker than that of the preceding, from which it is also distinguished by flexuous veins, proceeding from the base of the frond. It is annual, and fruits in summer. Towards the end of the season the fronds have become dark-coloured, and so destitute of gelatine that they do not adhere to paper. In this state it is very generally spotted with pretty *Lepraliæ.* In Ireland it grows to a great size, ten inches or so in length. It is rather a common plant in most places; but in Ayrshire it occurs less frequently. It may, however, generally be got in winter, in a young state, on the roots of *Halidrys* and *Lam. digitata.*

3. NITOPHYLLUM HILLIÆ, *Greville.*

Hab. In deep tidal pools. Rare. Annual. Summer and autumn. Miss Hill, Messrs. Rohloff, Hore, Cocks, Ralfs, Mrs. Griffiths, Dr. Jacob, and Dr. Arnott, England; Prof. Harvey, Valentia Island; Miss Turner, Jersey; Miss White, Scilly Islands and Isle of Portland.

It is obscurely veined at the base of the frond, which is of a thickish membranaceous substance, " resembling," says Mrs. Griffiths, " soft kid-leather." The dot-like granules are very small; the colour is rose-red; the smell, when fresh, is very disagreeable. See Pl. CLXIX. in Phyc. Brit., for a fine figure of it, of the natural size, and fruit, &c., magnified.

4. NITOPHYLLUM BONNEMAISONI, *Greville*.

This has been got by Dr. Greville in Bute; by the Rev. C. Clouston in Orkney; by Miss Ball, Miss Taylor, Professor Harvey, Mr. W. Thompson, in Ireland; by Mrs. Griffiths in England; by Miss White, Jersey. For figure and description distinguishing it from its congeners, see 'Phycologia Britannica,' Pl. XXIII.

5. NITOPHYLLUM GMELINI, *Greville*.

Hab. On rocks in the sea. Annual. Summer. Mrs. Griffiths, Miss Hill, Professor Walker Arnott, Miss White, Miss Cutler, Mrs. Gulson, in England; Miss Hutchins,

Fitch lith.

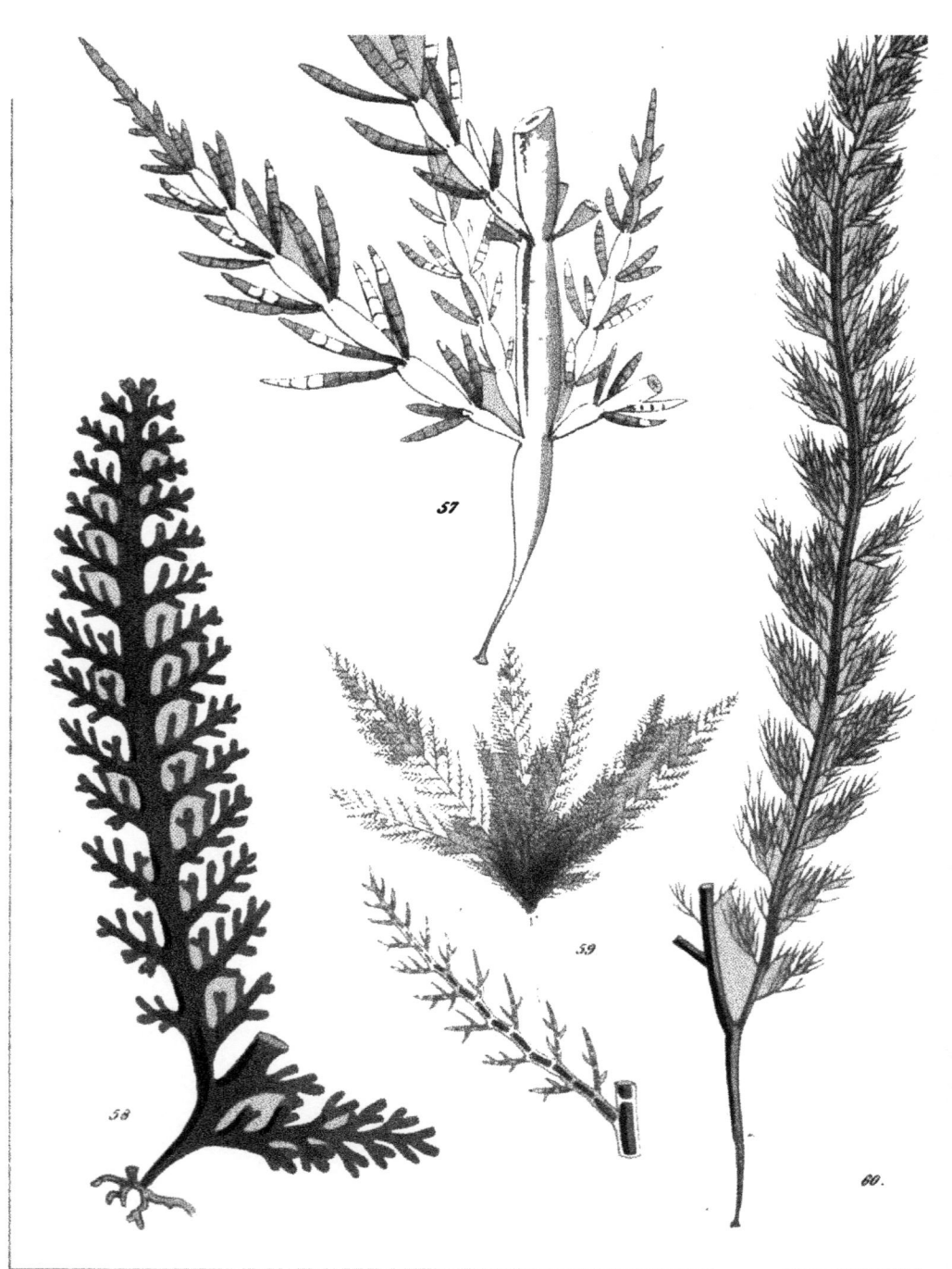

Mr. W. Thompson, Dr. Drummond, Mr. D. Moore, in Ireland; Major Martin at Ardrossan.

Substance membranaceous; colour a purplish rose-red.

6. NITOPHYLLUM VERSICOLOR, *Harvey*.

Hab. Found by Mrs. Griffiths and Miss Hill at Ilfracombe; by Miss White, Isle of Portland; not rare.

To Mrs. Griffiths it had long been known under the colloquial name of "Orange Dwarf," from its small size when compared with others; and from its turning from rosy pink to golden orange when exposed even to a shower of rain.

Family XV. CHONDRIEÆ.

"How often we forget all time, when lone
Admiring Nature's universal throne;
Her woods, her wilds, her waters,—the intense
Reply of hers to our intelligence!
Live not the stars and mountains? Are the waves
Without a spirit? Are the dropping caves
Without a feeling in their silent tears?
No, no. They woo and clasp us to their spheres,—
Dissolve this clog and clod of clay before
Its hour; and merge our souls in the great shore."

Genus LXXIV. BONNEMAISONIA, *Agardh*.

Gen. Char. Frond membranaceous, compressed, or plane,

filiform, much branched, the branches pectinate with distichous cilia. Fructification, sessile or pedicellate capsules, containing a cluster of pyriform (compound?) seeds, fixed by their base.— Named after Bonnemaison, a celebrated French Algologist.— *Greville.*

1. BONNEMAISONIA ASPARAGOIDES, *Agardh*. (Plate XII. fig. 45, a branch, natural size; and at the base, on the left, is the top of a branch, with capsules magnified.)

Hab. On submarine rocks. Annual. June to September. Found in England by Mr. Wigg, Mr. D. Turner, Mr. Stackhouse, Mrs. Griffiths, Miss Warren, Rev. Mr. Hore, Mr. Ralfs; in Ireland, Miss Hutchins, Dr. Drummond, Miss Gower, Professor Harvey, Mr. M'Calla; in Scotland, Major Martin, Ardrossan; Isabella L., at Saltcoats; Rev. M. Lambie, Southend, Argyleshire; dredged by Mr. Gourlie, Frith of Clyde, off Skelmorlie.

Dr. Greville (whose Pl. XII. in Alg. Brit. is very instructive) calls this an extremely elegant and beautiful plant. Professor Harvey describes it as "a highly beautiful species, and so unlike any other British Alga that it mus be recognised at a glance. The delicate *cilia* which border every part of the frond, and which are arranged with strict regularity, being always perfectly distichous, and placed *alternate* to each other, and *opposite* either to a capsule or to a branch, taken in connection with the cellular frond and brilliant colour, afford marks that cannot be mistaken." It is interesting to trace the history of beautiful plants, and it shows the rapid progress that the study of Algology is making, when we see that many which our most celebrated Algologists forty years ago spoke of as rare have been found to be not uncommon on many of our shores. This very plant, as the list of names shows, has been got in many parts of England, Scotland, and Ireland. At Southend and at Macrihanish Bay, in Argyleshire, it was so frequent last August, that D. L., jun., almost filled his vasculum in the course of an hour with beautiful specimens. During the summer and autumn of 1850 it was got in great abundance off Arran and Cumbraes, and on the coast of Ayrshire.

Mr. D. Turner says, "For the original discovery of this interesting *Fucus*, we are indebted to Mr. Wigg. It appears to be a species of remarkably unfrequent occurrence, since it is not known to exist beyond the limits of the British Isles,* and in the few places where it has been gathered has been far from plentiful. Seen floating in the water, nothing can be more elegantly feathery and delicate than its general

* It has since been found on several of the European shores, but not out of Europe.

appearance, the exquisite beauty of which those alone can appreciate who have had the opportunity of observing it recent, as whatever care may have been employed in the preservation of it, it is, nevertheless, in that state far inferior to what it was before it was dried." Mr. D. Turner's Pl. CI. is good, but in this faded state. I have, however, dried specimens of it whose colour is almost equal in splendour to the remarkably fine and correct figure given in 'Phycologia Britannica,' Pl. LI. The colour is a fine pellucid crimson.

Genus LXXV. LAURENCIA, *Lamour.*

Gen. Char. Frond cylindrical, filiform, between cartilaginous and gelatinous, mostly yellowish or purplish red. Fructification of two kinds: 1, ovate capsules, with a terminal pore, containing a cluster of stalked pear-shaped seeds, fixed by their base; 2, ternate granules, imbedded in the ramuli.—*Greville.*

1. LAURENCIA PINNATIFIDA, *Lamour.* (Plate XV. fig. 58, a portion of the frond, natural size.)

Hab. Rocks in sea. Annual. June to September.

This is one of the most common, and at the same time one of the most variable of our Algæ. It is found at high and low-water mark, and in deep water, its size and colour varying accordingly. It is very common on the coast of Ayrshire, where it ranges from an inch to upwards of six inches in length. Near high-water mark it is little more than an inch long, with a recumbent, curled, matted appearance, of a dark olive-colour. In rock-pools it is a little larger and yellower. At low-water mark it attains its full size, and is of a darker purplish colour. As in some states it does not adhere to paper, or stains the paper to which it adheres, it is occasionally treated to a moment's immersion in boiling water. This renders it more pliant, and less apt to stain the paper, but makes it paler in colour.

It is very interesting to look at the numerous varieties of this plant as figured by our most distinguished Algologists, Stackhouse, Turner, Greville, Harvey. The variety *Osmunda*, as figured by Stackhouse, is beautiful. Dr. Greville's figures in Pl. XIV. Alg. Brit., and Professor Harvey's in Pl. LV. Phyc. Brit., are very valuable. It is called *pepper-dulse*, and it certainly has, especially when young, a very pungent smell and peppery taste, and may form a very comfortable *quid* for an Icelander. It was formerly eaten in Scotland, but we rather think that now it is not at all used.

2. LAURENCIA CÆSPITOSA, *Lamour*.

Hab. On stones, &c., within tide-mark. Not uncommon. Intermediate between *L. pinnatifida* and *L. obtusa*.

3. LAURENCIA OBTUSA, *Lamour.*

Hab. On the larger Algæ. Annual. Summer and autumn.

This is common in England and Ireland. In Scotland, rather rare. It has been procured by Dr. Greville in Bute; by Mr. W. Thompson at Ballantrae; by Major Martin at Ardrossan; by D. L. at Portincross, Ayrshire, and in Arran and Kintyre pretty abundantly. It is very commonly found in rock-pools. When exposed to the sun, it becomes a dirty yellow; but when young, or growing in deep water, it is of a fine light-pink colour. It is from three to six inches in height. There is a beautiful characteristic figure in Phyc. Brit., Pl. CXLVIII.

4. LAURENCIA DASYPHYLLA, *Lamour.*

Hab. On rocks and stones in the sea. Annual. Summer. Frequent on the eastern and southern shores of England. Found in Ireland by Miss Hutchins, Mr. W. Thompson, Dr. Drummond; in Scotland, by Mr. Brodie, Lossiemouth; Dr. Greville, Bute; Major Martin, Ardrossan; in Arran, by Dr. Greville and D. L.; Miss White, Isle of Portland.

It is rarer in Scotland than the preceding, and it is a larger and more handsome plant; but its chief distinctions are the transverse striæ of the branches, and the tapering of the ramuli towards the base; colour pale pink, becoming yellow when in shallow water.

5. LAURENCIA TENUISSIMA, *Greville*.

Hab. In the sea, on rocks. Very rare. South of England and Ireland. Annual. Summer. Miss White, Isle of Portland and Jersey; fine and plentiful.

It differs from *L. dasyphylla* by the ramuli tapering both towards the base and apex. For good figures and descriptions, see Phyc. Brit., Pl. CXLVIII.

Genus LXXVI. CHRYSIMENIA, *J. Agardh*.

Gen. Char. Frond tubular, continuous (not constricted or jointed), filled with a watery juice, and traversed by longitudinal filaments; its scales composed of several rows of cells, the innermost of which are distended, and much elongated, the outer gradually smaller, and the ultimate very minute. Fructification of two kinds: 1, ovate or conical capsules (ceramidia), containing a dense mass of angular spores fixed to a central placenta; 2, triparted tetraspores immersed in the ramuli.—The name is from two Greek words signifying *golden membrane*, because the species acquires golden tints if long steeped in fresh water.—*Harvey*.

1. CHRYSIMENIA CLAVELLOSA, *J. Agardh*. (Plate XIV. fig. 56, a branch, natural size; and to the right at the base a small branch with capsules, magnified.)

Hab. On rocks, stones, shells, and Algæ at low-water mark and deeper. Annual. Spring and summer. We have got it on scallops dredged from deep water in Lamlash Bay. It is found on all the British shores. On the coast of Ayrshire it is in general rather rare; but in the summer of 1847 it was very abundant, being cast up by every tide. In 1850 it was got in considerable abundance in the west, but not fine. Fine specimens of it were gathered at Dunbar by Mrs. R. M. Stark.

It is a very variable plant, but beautiful in all its phases; some of them remarkably so, with fine arched branches, resembling the New Zealand *Chrysimenia secundata,* only our British plant is not secund. However, it even surpasses the New Zealander in colour, being a fine light pink, which it retains in drying. The figures in Hist. Fucorum are good, but those in Phyc. Brit., Pl. cxiv., are still better. *Chrysimenia* is distinguished from *Chylocladia* by the absence of internal diaphragms dividing the branches into distinct joints.

Genus LXXVII. CHYLOCLADIA, *Greville.*

Gen. Char. Frond (at least the branches) tubular, constricted at regular intervals, and divided by internal diaphragms into joints, filled with a watery juice, and traversed by a few longitudinal filaments; periphery composed of small polygonal cells. Fructification of two kinds, on distinct individuals: 1, spherical, ovate, or conical capsules (ceramidia), containing a tuft of wedge-shaped spores on a central placenta; 2, tripartite tetraspores immersed in the swollen branches near the apices.—Name from Greek words signifying *juice* and *branch.*—*Harvey.*

1. CHYLOCLADIA OVALIS, *Hooker.* (Plate VIII. fig. 31.)

Hab. On rocks and stones within tide-marks. Annual. Spring and summer. Frequent on the English and Irish shores. Rare in Scotland. Little Isles of Jura, Lightfoot; Lieut. Thomas and Dr. M'Bain, Papa Westra; we have it from Miss White, Jersey.

2. CHYLOCLADIA KALIFORMIS, *Hooker.* (Plate XV. fig. 57, a portion of the frond.)

Hab. In the sea on rocks and on Algæ, generally in deep water, but occasionally in shallow water at mid-tide mark. Annual. June to September. It is frequent in England and Ireland, and not rare in Scotland, and is much more common in Arran than on the Ayrshire coast.

It is a handsome plant when dried, though when fresh from the sea it has a clumsy appearance, as in general it is in a bunch of a dozen branches, nearly a foot in length, subgelatinous, and filled with watery juice; when fresh from deep water the colour is purple. When partly decomposed, it is at times singularly beautiful, red, purple, and yellow and white, and green tints being richly mingled. More frequently, however, it is a dull bluish-purple when driven ashore. There is a magnificent figure of its normal appearance in Phyc. Brit., Pl. cxlv.

* 3. CHYLOCLADIA REFLEXA, *Chauv*.

Hab. On rocks in the sea near low-water mark. Very rare. Found by Miss Amelia Griffiths near Ilfracombe, and by Mr. Ralfs in Cornwall, from whom we have good specimens. It is chiefly distinguished from *C. kaliformis* by its creeping habit and small size.

4. CHYLOCLADIA PARVULA, *Hooker*.

Hab. On large Algæ. Annual. Summer and autumn. England, Mrs. Griffiths, Mr. Borrer; Ireland, Miss Hutchins, Mr. Templeton, Mr. D. Moore; at Saltcoats, floating in the sea, by Mrs. R. M. Stark; island of Cumbrae, Rev. G. Laing.

Colour, a fine, fugitive pinky red. Distinguished from *C. kaliformis* by the ramification, the uniformly short joints, and the shape of the capsules, which are ovate. See Phyc. Brit., Pl. ccx.

5. CHYLOCLADIA ARTICULATA, *Hooker*. (Plate XIV.

fig. 53, portion of plant, natural size, and the base, on the left, is a small portion with capsules, magnified.)

Hab. On rocks, and on the larger Algæ. Annual. Summer.

Common on all our shores, creeping on the steep sides of perpendicular rocks near low-water mark. Irish specimens are large, sometimes a foot in length. In the west of Scotland they are seldom above five inches in length, and more generally not above three inches. The smaller specimens are like *Catenella Opuntia*. Colour reddish-purple; substance membranaceous, filled with watery gelatine.

Family XVI. CORALLINEÆ.

"Involved in sea-wrack here you find a race,
Which Science, doubting, knows not where to place."—*Crabbe*.

"Corallinas ad *Regnum Animale* pertinere ex substantiâ earum calcareâ constat, cum omnem calcem Animalium esse productum verissimum sit."—*Linnæus*.

"Mihi verò totum hocce genus *Botanicis* relinquendum videtur."—*Pallas*.

"Heu quæ nunc tellus—quæ me æquora possint
Accipere——?
Cui neque apud Danaos usquam locus."—*Virgil*.

Genus LXXVIII. CORALLINA, *Linnæus*.

Gen. Char. Frond filiform, articulated, branched (mostly pinnate), coated with a calcareous deposit. Fructification, turbinate or obovate, mostly terminal ceramidia, pierced at the apex by a minute pore, and containing a tuft of erect, pyriform, or club-shaped, tranversely parted tetraspores.—The name is from their resemblance to coral.—*Harvey.*

1. CORALLINA OFFICINALIS, *Linnæus.* (Plate XIV. fig. 54, natural size, and on the left, a portion of a branch with ceramidia, magnified.)

Hab. On rocks between tide-marks. Perennial. Winter and spring. Abundant on all our rocky shores.

Dr. Johnston, in his well-known 'History of British Sponges and Corallines,' says, respecting this species, "It appears first in the guise of a thin, circular, calcareous patch, of a purplish colour, and in this state is common on almost every object that grows between tide-marks." It has been described as *Millepora lichenoides,* while its earlier states constitute Lamouroux's various species of *Melobesia.* A very beautiful white light is produced by holding a piece of *C. officinalis* close to the flame of a candle. It was once believed to have very powerful vermifuge virtues. The colour is a dark purple, soon fading on exposure. Though kept dry for years it continues to emit an unpleasant smell.

Professor Harvey says, "The question of the vegetable

nature of *Corallines* may now be considered as finally set at rest by the researches of Kützing, Philippi, and Decaisne. Whoever macerates a portion of one of these stony vegetables in acid, till the lime it contains be dissolved, will find that he has a structure of a totally different nature from that of any zoophyte, while it is perfectly analogous to that of many Algæ." In Pl. ccxxii. of Phyc. Brit. there is a figure of a portion of the frond after maceration in acids.

2. CORALLINA ELONGATA, *Ellis and Solander*.

Vide 'Phycologia Britannica;' see also Dr. Johnston's 'History of British Sponges and Corallines,' page 221.

3. CORALLINA SQUAMATA, *Ellis and Solander*.

Hab. On submarine rocks. Perennial. Summer. South coast of England, Ellis, &c.; West of Ireland, Professor Harvey; Youghal, Miss Ball. There is no Scottish habitat given, but we have seen it on the Ayrshire coast; and we have it also from Australia.

Genus LXXIX. JANIA, *Lamouroux*.

Gen. Char. Frond slender, branched in a dichotomous manner; the joints cylindrical; the crust calcareous, unporous; the axis subcartilaginous, solid, constricted at intervals corresponding to the articulations of the crust. Capsular swellings produced in the axis of the branches containing several granules.—*Johnston*.

1. J'ania rubens, *Lamouroux*.

Hab. Generally on the smaller Algæ. Dr. Johnston says that it is not found on the coast of Berwickshire, and that it is rare on the eastern shores of Scotland. It is frequent on the English and Irish shores. It is rare on the coast of Ayrshire about Ardrossan and Saltcoats, but not rare a little further north at Portincross, and very abundant more to the south at Ballantrae, where I have gathered a score of specimens in a quarter of an hour. It is frequently found in the island of Arran.

It grows in dense tufts from an inch to two inches in height. Colour pale, but, when exposed to the sun, it becomes whitish or at times greenish. Dr. Johnston says that the structure of the axis is decidedly vegetable; and Dr. Harvey is of the same opinion. For an excellent figure of it, natural size, and also for a magnified figure after the crust has been removed by acid, see Johnston's ' History of British Sponges and Corallines,' and ' Phycologia Britannica,' Pl. CCLII.

Genus LXXX. MELOBESIA, *Lamouroux*.

Gen. Char. Frond attached or free, either flattened, orbicular, sinuated, or irregularly lobed, or cylindrical or branched (never articulated), coated with a calcareous deposit. Fructification, conical, sessile capsules (ceramidia) scattered over the surface of the frond, and containing a tuft of transversely parted oblong tetraspores.—Name from one of the *Sea Nymphs of Hesiod.*— *Harvey.*

1. MELOBESIA POLYMORPHA, *Linnæus.*

Hab. On submarine rocks and in quiet bays. We have dredged it in Lamlash Bay, where there are extensive beds of it, at the depth of several fathoms. Similar beds are found at Rothesay, and in Loch Fyne.

It is very hard and very diversified in form, as the specific name implies. Ray says that it is dredged out of Falmouth Harbour to manure their lands in Cornwall; and Mr. W. Thompson informs us that it is dredged in Bantry Bay for the same purpose. From Professor John Fleming we learn that it is so abundant in Orkney as to warrant the conclusion that it might be advantageously employed for agricultural purposes and for building, especially as limestone is scarce in Orkney, and generally of bad quality. Dr. Walker, in his Essays, says, " Of the cathedral of Icolmkill (Iona), the cement is so strong that it is easier to break the stones than to force them asunder. It is of lime that has been calcined from sea-shells, and formed into a very gross mortar, with coarse gravel, in a large proportion, and a great quantity of the fragments of *white coral,* which abounds upon the shores of the island." The colour of this coralline is generally white when it lies bleached on the shore, but when newly dredged in Lamlash Bay it is of a reddish purple. At first I was disappointed when the dredge came up full of this Millepore. I soon learned, however, to hail its appearance, for on examining it carefully, handfull by handfull, I found many precious things intermingled with the coral. One of these was *Lima tenera,* which, like those persons who built the cathedral of Icolmkill, employs much of this coral in forming its habitation.*

* The most interesting, though not the rarest thing we got was *Lima tenera,* Turt. I had in my cabinet specimens of this pretty bivalve, and I had admired the beauty and elegance of the shell, but hitherto I had been unacquainted with the life and manners of its inhabitant. Mr. and Miss Alder, of Newcastle, who were of the party, had got it in the same kind of coral at Rothsay; so that when Miss Alder got a cluster of this coral, brought up by the dredge, cohering in a mass, she exclaimed, "Oh, here is the *Lima's* nest!" and, breaking it up, the *Lima* was found snug in the midst of it. The coral nest is curiously constructed, and remarkably well fitted to be a safe residence for this beautiful animal. The fragile shell does not nearly cover the mollusk,—the most delicate part of it, a beautiful orange fringe-work, being altogether outside of the shell. Had it no extra protection, the half-exposed animal would be a tempting mouthfull, quite a *bonne bouche,* to some prowling haddock or whiting; but He who tempers the wind to the shorn lamb, teaches this little creature, which He has so elegantly formed, curious arts of self-preservation. It is not content with hiding itself among

61.

62.

63.

64.

2. MELOBESIA AGARICIFORMIS, *Lam.*

See Dr. Johnston's 'History of British Sponges and Corallines,' page 241, and Professor Harvey's 'Phycologia Britannica,' Pl. LXXIII.

the loose coral,—for the first rude wave might lay it naked and bare: it becomes a marine mason, and builds a place of abode; it chooses to dwell in a coral grotto, but in constructing this grotto it shows that it is not only a mason, but a rope-spinner, and a tapestry-weaver, and a plasterer. Were it merely a mason, it would be no easy matter to cause the polymorphous coral to cohere. Cordage, then, is necessary to bind together the angular fragments of the coral; and this cordage it spins, but its mode of spinning it is one of the secrets of the deep. Somehow or another, though it has no hands, it contrives to intertwine this yarn which it forms, among the numerous bits of coral, so as firmly to bind a handfull of it together. Externally, this habitation is rough, and therefore better fitted to elude or to ward off enemies; but, though rough externally, all is smooth and lubricous within, for the yarn is woven into a lining of tapestry, and the interstices are filled up with fine slime, so that it is smooth plaster-work, not unlike the patent Intonaco of my ingenious friend, Mrs. Marshall. Not being intended, however, like her valuable composition, to keep out damp, or to bid defiance to fire, while the intertwining cordage keeps the coral walls together, the fine tapestry, mixed with smooth and soft plaster, covers all asperities, so that there is nothing to injure the delicate fringed appendages of the enclosed animal. Tapestry, as a covering for walls, was once the proud and costly ornament of regal apartments; but ancient though the art was, I shall answer for it that our little marine artizan took no hint from the Gobelins, or from the workmen of Arras, or from those of Athens, or even from the earliest *tapissiers* of the East. I doubt not that from the time that Noah's Ark rested on the mountain of Ararat, the forefathers of these beautiful little *Limas* have been

3. MELOBESIA FASCICULATA, *Harvey.*

Hab. Found by Mr. M'Calla on the sandy bottom of the sea, in 4-5 fathoms water.

It is from one to three inches in diameter, irregularly lobed. I have a specimen from Loch Ryan which perfectly corresponds with the first figure of it, given in Pl. LXXIV. of Phyc. Brit.

After the description of it, Professor Harvey adds, " I have mentioned that the vegetable nature of the corallines is now distinctly proved. The question still remains whether *Melobesiæ* are independent vegetables, or whether they be constructing their coral cottages, and lining them with well-wrought tapestry in the peaceful bay of Lamlash.

When the *Lima* is taken out of its nest, and put into a jar of sea-water, it is one of the most beautiful marine animals you can look upon. The shell is elegant, the animal within the shell is beautiful, and the orange fringe-work outside of the shell is highly ornamental. Instead of being sluggish, it swims about with great vigour. Its mode of swimming is the same as that of the scallop. It opens its valves, and, suddenly shutting them, expels the water, so that it is impelled onwards or upwards; and when the impulse thus given is spent, it repeats the operation, and thus moves forward by a succession of jerks, or jumps. When moving through the water in this way, the reddish fringe-work is like the tail of a fiery comet. The filaments of the fringe may, for anything we know, be useful in catching their prey; they are very easily broken off, and it is remarkable that they seem to live for many hours after they are detached, twisting themselves about in a vermicular manner.—*Excursions to Arran, by D. L.,* p. 319.

merely amorphous states of the common *Corallina officinalis*. This latter is the view of the subject advocated by Dr. Johnston, whose opinion—founded on observation, and as the opinion of an accomplished naturalist who has paid much attention to the lower tribes of animals, and is familiar with variations in form among sponges, nearly as wild as this would be—must not be hastily condemned." Though he does not feel warranted to give a direct negative, he is not disposed to agree with Dr. Johnston's views on this point, and he assigns his reasons.

> "The Ocean heaves resistlessly
>   And pours his glittering treasure forth;
> His waves, the priesthood of the sea,
>   Kneel on the shell-gemmed earth,
> And there emit a hollow sound
>   As if they murmur'd praise and prayer:
> On every side 'tis holy ground—
>   All nature worships there!"—*Vedder.*

Family XVII. RHODOMELEÆ.

"The Ocean old hath my deep reverence,
And I could watch it ever:—when it sleeps,
And its hushed waves but throb at intervals,
Like some fair infant's breath in sad repose,—
'Tis strangely sweet to gaze;—or when it starts
At voice of torturing storm, and, like mad age,
Tosses its hoar hair on the raving wind,
'Tis wild delight to watch it. But I love
To see it gently playing on loose rocks,
Lifting the idle sea-weed carelessly;—
Or hear it in some dreary cavern muttering
A solitary legend of old times."—*T. H. Reynolds.*

Genus LXXXI. ODONTHALIA, *Lyngbye.*

Gen. Char. Frond plane, between membranaceous and cartilaginous, dark venous red, with an imperfect or obsolete midrib, and alternately toothed at the margin. Fructification, marginal, axillary, or in the teeth: 1, capsules (ceramidia) containing pear-shaped seeds, fixed by their base; 2, slender processes (stichidia) containing ternate granules.—*Greville.*

1. ODONTHALIA DENTATA, *Lyngbye.*

Hab. On rocks in the sea. Perennial. Fruiting in early winter. Filey, Yorkshire, by Mrs. Gatty.

This is a northern species, not being got in England south of Durham; and it is confined also in Ireland to the north.

It is most abundant in Scotland; very frequent on the coast of Ayrshire, where we have a variety of it richer than any I have seen figured. The fronds of this variety are rather more limber than the common kinds, and the branches and denticulations are smaller and more numerous, and more closely arranged. In November, *Odonthalia* is very frequent on the shore, finely dotted with fruit, which is very observable, as the frond at this season is dark-coloured, and the fresh ceramidia and stichidia are reddish-purple. In the early part of the season, April and May, the young light-brown fronds are intermingled with the old dark fronds. The young fronds adhere well to paper; the old have become rigid, and do not adhere unless coaxed with a little isinglass, or subdued by a hasty ducking in hot water. Dr. Greville's Plate XIII. in Alg. Brit., and Professor Harvey's Plate XXXIV. in Phyc. Brit., gave much insight as to the structure and fructification. The colour of Plate XXXIV. is bluer than the plant is with us. Mr. D. Turner's is too brown.

Genus LXXXII. RHODOMELA, *Agardh*.

Gen. Char. Frond cylindrical or compressed, filiform, much branched, coriaceo-cartilaginous (the apex sometimes involute).

Fructification: 1, subglobose capsules, containing free pear-shaped seeds: 2, pod-like receptacles, with imbedded ternate granules.—*Greville.*

1. RHODOMELA SUBFUSCA, *Agardh.*

Hab. In the sea, on rocks and on other Algæ. Perennial. Fruiting from the beginning of spring till the beginning of winter, when it bears stichidia. Very common on the English, Irish, and Scottish shores.

It is a very variable plant, in spring beautifully tufted, in summer coarse and bushy, in winter ragged, rigid, and ugly; and yet even then it is far from being without interest, as it is then occasionally ornamented with fresh-coloured stichidia on the old black branches. The colour in summer is a dull reddish-brown, which becomes black in drying. I have some fine black specimens from Staffa and Iona. Except in its young tufted state, it does not adhere well to paper.

2. RHODOMELA LYCOPODIOIDES, *Agardh.*

Hab. On rocks in the sea, and on *Laminaria digitata.* Perennial. Fruiting in spring and summer. Its capsular fruit is more globose than that of the preceding.

This is a northern species, frequent in the north of England and Ireland, and found in most of the Scottish shores. In spring it is procured in considerable abundance on the coast of Ayrshire. The winter and early summer states of this plant are exceedingly different. In winter, when found with perhaps a dozen black branches, more than a foot in length, from one root, closely invested with the numerous stumps of summer branchlets, it might pass for a cluster of black wolves' tails, rather than *wolves' feet*, as the specific name would lead us to expect, unless the black feet carry the black legs along with them. In early summer it sends forth along the old stem abundance of young branches, which are beautified with numerous feathery tufts of ramuli, causing it greatly to resemble *Polysiphonia Brodiæi*. The substance in winter is cartilaginous and rigid, and does not adhere to paper in drying; the colour in that state is very dark. In early summer the branches are quite pliant and rather flaccid, adhere closely to paper, and form a beautiful specimen, of a light brownish-purple colour. In some specimens, collected on the coast of county Down, the frond is twenty inches long, and the lateral branches from six to fourteen: some of the Scotch specimens are equally large. The figure in Hist. Fucorum is good, in the winter state of this plant, though too light in colour. Plate L. in Phyc. Brit. is good and instructive.

GENUS LXXXIII. BOSTRYCHIA, *Mont.*

Gen. Char. Frond dull purple, filiform, much branched, inarticulated, dotted; traversed by a jointed tube surrounded by one or more concentric layers of oblong, coloured cells, which are gradually shorter towards the circumference; the surface cells quadrate. Fructification of two kinds on distinct individuals: 1, "lateral capsules" (ceramidia), *Roth.*; 2, tetraspores, contained in terminal, lanceolate pods.—The name is from a Greek word signifying *a curl of hair*, or *ringlet*.—*Harvey.*

1. BOSTRYCHIA SCORPIOIDES, *Mont.* (Plate XI. fig. 41, branch of the natural size; and small portion of a branch, magnified.)

Hab. On muddy sea-shores, near high-water mark; at the estuaries of rivers; in salt-water ditches and marshes, and even adhering to the roots of flowering-plants. Annual. Summer. In England and Ireland, but, in so far as we know, not yet observed in Scotland.

When revising this, and wishing to be better acquainted with this interesting plant, of which I had only one specimen from Mrs. Griffiths, a packet arrived from Miss Cutler, containing more than a score of fresh undried specimens, got at the mouth of the river. This was a treat, for which I felt much indebted to that kind and excellent lady, who is so well known to Algologists. The finest plants she found in a marsh, growing on *Atriplex portulacoides.*

Genus LXXXIV. RYTIPHLÆA, *Agardh.*

Gen. Char. Frond filiform or compressed, pinnate, transversely striate, reticulated; the axis articulated, composed of a circle of large, tubular, elongated cells (siphons) surrounding a central cell; the periphery of several rows of minute, irregular, coloured cellules. Fructification of two kinds, on distinct individuals: 1, ovate capsules (ceramidia) containing a tuft of pear-shaped spores; 2, tetraspores, contained in minute lanceolate receptacles (stichidia) in a double row.—The name is from two Greek words signifying a *wrinkle* and *bark*, because the surface is transversely wrinkled or striate.—*Harvey.*

1. RYTIPHLÆA PINNASTROIDES, *Agardh.*

Hab. On submarine rocks near low-water mark. Perennial. Winter. In several places on the shores of the south of England. Rev. J. Z. Edwards, near Axminster. By Miss White and Miss Turner in Jersey.

Found neither in Scotland nor Ireland. Substance cartilaginous, not adhering to paper; colour, a dark reddish-brown, becoming black in drying. For structure, fruit, &c., see Plate LXXXV. of 'Phycologia Britannica.'

2. RYTIPHLÆA COMPLANATA, *Agardh.*

Hab. In rock-pools, among *Corallina officinalis.* Perennial? Summer. Very rare. Discovered in Bantry Bay by Miss Hutchins. Obtained by Professor Harvey, Ireland; by the Rev. W. S. Hore, Professor Walker Arnott, Dr. Jacob, in England. See fine figure, Phyc. Brit., Pl. CLXX.

3. RYTIPHLÆA THUYOIDES, *Harvey.* (Plate XV. fig. 59, plant of natural size; and part of a branch, magnified.)

Hab. Rocks in the sea, rock-pools. Perennial. Summer and autumn. Bantry Bay, Miss Hutchins; Prof. Harvey, Miltown Malbay; Mrs. Griffiths and Mr. Ralfs, Devonshire; Mr. W. Thompson, Ballantrae; D. L., Portincross; D. L., jun., Arran.

It is a variable plant, some specimens resembling *Polysiphonia nigrescens,* and others approaching *Polysiphonia fruticulosa.* It often bears yellow antheridia in summer; colour of the plant, a brownish-red, becoming darker in drying.

4. RYTIPHLÆA FRUTICULOSA, *Harvey.*

Hab. On rocks in the sea; sometimes in rock-pools, and at other times on sand-covered rocks. Perennial. Summer and autumn. Frequent on the southern shores of England. Got in Ireland by Miss Hutchins, Mr. W. Thompson, Mr.

D. Moore; in Scotland, by Capt. Carmichael, at Appin; by Miss Ramsay, Spring Bank, Arran; by D. L., Corrie, Arran; by D. L., jun., Corriegils, Arran. I know only of one place where it is got on the coast of Ayrshire,—in the little harbour at Portincross, where it was first found by Mrs. Ovens and Major Martin.

Though cartilaginous, it adheres to paper, and forms a beautiful specimen. Colour, brownish, becoming black in drying.

Genus LXXXV. POLYSIPHONIA, *Greville*.

Gen. Char. Frond filamentous, partially or generally articulate; joints longitudinally striate, composed internally of parallel tubes, or elongated cellules. Fructification twofold, on distinct plants: 1, ovate capsules (ceramidia) furnished with a terminal pore, and containing a mass of pear-shaped seeds; 2, tetraspores imbedded in swollen branchlets.—The name is from two Greek words signifying *many tubes.—Harvey.*

1. POLYSIPHONIA PARASITICA, *Greville*. (Plate XII. fig. 47, plant of the natural size; to the right, a portion of a pinnule, and to the left, a portion of a pinnule with a capsule, both magnified.)

Hab. Parasitical on the larger Algæ, and more frequently

on *Melobesia* on the steep sides of rock-pools. Got also by dredging in from four to fifteen fathoms water. Found on all the British coasts, but regarded as rather rare. Mrs. Griffiths, Devonshire; Dr. Greville, Loch Ryan; Major Martin, Ardrossan; D. L., Saltcoats. Many fine specimens have been procured in the island of Arran, where it was discovered by Isabella L., growing in abundance in rock-pools on *Melobesia*. The richest habitat, however, is at Portincross, betwixt Ardrossan and Largs, where in August it is abundant among *rejectamenta*, in the little creek that forms the harbour, being drifted from deep water.

Professor Harvey says that our Ayrshire specimens are by much the finest he has seen. The numerous branches are at times about three inches in length. It is often got with capsular fruit, and the capsules are larger in proportion to the stem. At times, also, there are spurious capsules,— knobs formed by the stinting of the ramuli. Not unfrequently, also, there are clusters of short branches matted together on various places of the frond,—miniature resemblances of those bunches of twigs like birds' nests often seen on birch-trees.

But the most interesting specimens, though the most minute, were got by me after the storms of December 1848, creeping on the roots of *Halidrys siliquosa*. Instead of growing upright, as they usually do, they were repent, not only fastened by the radical roots (if we may employ this tautological expression), but throwing out roots at intervals from the stem and from the branchlets, and so firmly did they adhere by these that it took repeated tugs to disengage them, and you heard a crackling as if you had been slitting out the stitches of an old garment. In this way they lay quite unaffected by any storm that did not uproot the *Halidrys*. The largest frond I observed of this creeping variety was not above an inch in length; it was much more compressed than the erect kind, and, being a darker purple, it bore a striking resemblance to *Jungermannia complanata*, which creeps on the bark of trees. I found it also in this state on the roots of *Laminaria digitata*, but *Halidrys* was evidently the favourite. Creeping in this way, it is much better entitled to the specific name *parasitica*, as it resembles ivy, putting out its tendrils to cling tenaciously to trees.

The colour is reddish-brown, becoming darker in drying. There is a beautiful figure of it in 'Phycologia Britannica,' Pl. CXLVII.

2. POLYSIPHONIA SUBULIFERA, *Harvey*.

Hab. In the sea. Rare. Perennial? Found by Mrs.

Griffiths, Torquay; Capt. Carmichael, Appin; by Professor Harvey (who kindly sent me specimens) in Ireland; found also on the Irish shores by Mr. Templeton and by Mr. M'Calla; by the latter abundantly at Roundstone.

It has a very thorny habit, as the specific name implies. To the naked eye it bears some resemblance to young specimens of *Rytiphlæa fruticulosa*, but it is distinctly jointed and more flaccid. See a good figure of it in Phyc. Brit., Pl. CCXXXVII.

3. POLYSIPHONIA ATRO-RUBESCENS, *Greville*.

Hab. On stones and rocks in the sea, near low-water mark. Annual. Summer and autumn. Not uncommon.

It has borne a good many different names,—*Conferva nigra, Hutchinsia denudata, Polysiphonia badia, P. Agardhiana*, and finally *P. atro-rubescens*: the specific name was given by Dillwyn. It is got on the coast of Ayrshire. One of the best distinguishing marks, though not constant, is the spiral curving of the tubes. Colour, a dark red, becoming brownish; the substance is somewhat rigid, so that it does not always adhere to paper. See in 'Phycologia Britannica,' Pl. CLXXII., a very good figure, natural size, and portions of the stem and fruit magnified.

4. POLYSIPHONIA SPINULOSA, *Greville*.

Hab. Torquay, Mrs. Griffiths; Appin, Capt. Carmichael. Extremely rare. A good figure of it is given by Dr. Greville in his 'Cryptogamic Flora.'

5. POLYSIPHONIA NIGRESCENS, *Greville*.

Hab. On rocks, &c., in the sea. Perennial. Summer.

Fronds tufted, and from six to eight inches high; filaments robust, rigid, and generally rough below, with broken branches, much branched and bushy above; articulations short, capsules ovate; colour dark-brown, becoming black in drying. In a young state it is often very beautifully tufted, and of a purplish-pink colour, which it retains in drying.

6. POLYSIPHONIA FURCELLATA, *Harvey*.

This is a very rare species. Mrs. Griffiths found it floating in the sea at Sidmouth. It was obtained also by Mr. M'Calla near Carrickfergus in 1846. It is between *Polysiphonia fastigiata* and *Polysiphonia nigrescens*, but different from both. See Phyc. Brit., Pl. VII.

7. POLYSIPHONIA FASTIGIATA, *Greville*.

Hab. On *Fucus nodosus* and *F. vesiculosus,* especially the former, forming globose, dense bushy tufts, of a brown or at times a yellowish colour. It is very common; but, common as it is, I have a liking for it, as it was the first of the small Algæ I knew by name. I had been corresponding with Professor John Fleming, of Edinburgh, respecting shells and zoophytes, and, meeting with this sea-weed in such abundance, I wished to know its name, and sent a specimen to him, which he returned with the name it then bore; so that I am indebted to him for setting me agoing in Algology as in some other departments of natural science. It is common at all seasons on all our shores. Many a little mollusk nestles among its tufted branches. In summer its tips are often yellow with *antheridia*. It adheres well to paper in the early part of the season, and makes a fine specimen.

8. *P. Richardsoni,* Hooker.   11. *P. Griffithsiana,* Harv.
9. — *Carmichaeliana,* Harv.   12. — *stricta,* Greville.
10. — *variegata,* Agardh.   13. — *Grevillii,* Harv.

14. POLYSIPHONIA OBSCURA, *Agardh.*

Jersey, Miss White; Sidmouth, Rev. R. Creswell. Plant from a quarter to half an inch in height. Colour a dark red brown.

15. POLYSIPHONIA BRODIÆI, *Greville.* (Plate XV. fig. 60, branch, natural size, and branchlet with capsules, magnified.)

Hab. On rocks and corallines near low-water mark. Annual. Summer. Common in Scotland on rocky shores; also in England and Ireland.

Frequent as it is now known to be, it remained unnoticed

65.

66.

67.

68.

by-naturalists till it was sent by Mr. Brodie, of Brodie, to Mr. Dillwyn about forty years ago, who dedicated it to Mr. Brodie. It was then ranked as a *Conferva*. " *C. Brodiæi*," says Mr. Dillwyn, " is among the most magnificent of the genus, often extending to a foot and a half or two feet in length, and pushing forth, from a discoid base, several main filaments as thick as small twine, and of a blackish-purple colour. These are beset with scattered branches of uncertain length. Along the branches, at irregular intervals, clusters of slender ramuli are disposed." The colour is generally dark brownish-purple; the substance is soft, soon decomposing in fresh water. There is a fine figure of it, and the different kinds of fruit, in Phyc. Brit., Pl. cxcv. The plate in Dillwyn's 'British Confervæ' is both good and interesting, for the drawing from which it was taken was made by Miss Hutchins, and communicated by her to Mr. D. Turner, who sent it to Mr. Dillwyn.

16. POLYSIPHONIA FIBRILLOSA, *Greville*.

Hab. Rocks in the sea. Annual. Summer.

This is a pretty species, common in England, Scotland, and Ireland. It is large, the frond being from 6–10 inches, the main stem nearly half a line in diameter; the branches in the upper part clothed with slender, finely divided ramuli, at times straw-coloured, but more generally rosy, becoming purple when dried. Mr. Borrer thinks that it resembles *Pol. byssoides*; Mrs. Griffiths considers it as more like *Pol. Brodiæi*. To us it seems, in a young fresh rosy state, to come nearer *Polysiphonia elongella*, but it loses much of the resemblance when it is dried. It is very common in the island of Arran, and not uncommon on the coast of Ayrshire.

17. POLYSIPHONIA VIOLACEA, *Greville*.

Hab. On rocks and stones and Algæ near low-water mark. Annual. May and June. First discovered in Britain by Mrs. Griffiths; Falmouth, Miss Warren; Carnarvon, Mr. Ralfs; Plymouth, Mr. Rohloff; Belfast Lough, Dr. Drummond; Howth, Miss Gower; Kerry, Mr. Andrews; Roundstone, by Mr. M'Calla. Got occasionally on the coast of Ayrshire. I have some fine specimens from Mrs. Griffiths and Mr. Ralfs. See a beautiful figure of it in 'Phycologia Britannica,' Pl. ccix.

18. POLYSIPHONIA FIBRATA, *Harvey*.

Hab. On rocks, shells, and Algæ, near low-water mark. Annual. Summer and autumn. Common.

It is articulated throughout, and the joints marked with two striæ. At Ardrossan it is found abundantly on *Chorda filum* at the mouth of the harbour, and, prepared when quite fresh, it makes a beautiful specimen. In fresh water it very soon decomposes. There is a fine figure of it in Phyc. Brit., Pl. CCVIII., and magnified specimens of three kinds of fruit, *ceramidia*, *tetraspores*, and *antheridia*, the last of which, viz., antheridia, are often clustered round the tips of the branches, " crowning every branchlet with a tuft of golden fruit."

19. POLYSIPHONIA PULVINATA, *Spreng.*

Hab. On rocks in the sea, and in rock-pools. Annual. Mrs. Griffiths, Torbay; Land's End, Mr. Ralfs; Miltown Malbay, Professor Harvey; Miss Gower, Balbriggan; Mr. D. Moore, Port Stewart; Ardrossan, Major Martin; Saltcoats, D. L., jun.

It resembles *Polysiphonia urceolata* in miniature. It is not only smaller but softer, and the branchlets are more closely crowded. See figure in Phyc. Brit., Pl. CII.

20. POLYSIPHONIA URCEOLATA, *Greville.*

Hab. On rocks and on the stems of *Laminaria digitata*. Annual. Summer. Growing on *Laminaria* it becomes rigid when full grown, and does not adhere to paper in drying. What grows on rock, is with us less rigid, darker coloured, and decomposes more readily in fresh water.

When growing on *Laminaria* the branchlets are often very squarrose, and in this state it has been called *P. patens*, but it is only a variety. See 'Phycologia Britannica,' Pl. CLXVII.

21. POLYSIPHONIA FORMOSA, *Suhr*.

Hab. On rocks and Algæ. Annual. Summer. Found in England by Mrs. Griffiths, Mrs. Wyatt, Rev. W. S. Hore, Dr. Cocks, Dr. Jacob, Rev. R. Cresswell; in Ireland by Mr. W. Thompson, Miss Ball, Mr. Moore, Mr. M'Calla; in Scotland, by Suhr, Orkney; Dr. Greville, Bute; Major Martin, Ardrossan; Miss M'Leish, Portincross.

On the coast of Ayrshire it has chiefly been found growing on *Halidrys siliquosa*. It is a beautiful plant, much resembling *Polysiphonia urceolata*, but it is more graceful and delicate, and its colour is more rosy.

22. POLYSIPHONIA ELONGATA, *Greville*.

Hab. In the sea, on stones, shells, and Algæ. Biennial. Spring and summer.

Stems robust, cartilaginous; branches beset at the tips with slender tufted ramuli, which are attenuated at the base. It is very common and very variable. In its winter state it is stout and rigid, destitute of ramuli, the branches jointed like lobsters' horns. In early summer, when the capillary branches in rich tufts adorn the tips of the branches, it is a beautiful plant, especially when the tufts are of a fine rosy-red colour.

23. POLYSIPHONIA ELONGELLA, *Harvey*. (Plate XIV. fig. 55, portion of the frond, natural size; *a*, branchlet with capsules; *b*, a branchlet with tetraspores, both magnified.)

Hab. On rocks and Algæ; rather rare. Biennial. Spring and summer. England, Mrs. Griffiths, Miss Cutler, Rev. W. S. Hore; Ireland, Miss Ball, Miss Gower, Mrs. Apjohn, Mr. W. Thompson, Dr. Drummond, Mr. D. Moore, Mr. M'Calla; Orkney, Rev. Mr. Pollexfen; Ardrossan, Major Martin; D. L., Saltcoats; by Miss H. M. White, Jersey.

This is one of the loveliest of our marine Algæ. In early summer, when arrayed in its new attire with rich tufts of crimson ramuli, it is extremely attractive. These crimson tufts form a good contrast with the darker hue of the stems. The summer specimens we have got on the coast of Ayrshire are more beautiful than any we have seen, even from Devonshire. In winter it is rigid, and quite in undress. "The winter and summer aspects of a deciduous tree are not more different from each other than are specimens of this beautiful plant collected at opposite seasons." In winter the abundant pencils of rosy-red ramuli have all fallen away, and the rigid divaricating branches remain in their nakedness. It seems to be one of those that come by fits and starts. In the early part of the summer of 1847, for nearly a month, it was frequent on the shore at Ardrossan and Saltcoats at almost every tide; whereas in the summer of 1848 very few specimens were seen. In the summer of 1850 some plants were obtained, but not in beauty. Professor Harvey says, in Phyc. Brit., Pl. CXLVI., that though it resembles *Polysiphonia elongata* in miniature, it may be readily known from it by the pellucid articulations visible in all parts of the plant, and by the ramuli not tapering to the base; and in his Manual he tells us that it may easily be distinguished by the distinctly-jointed branches, and by the parallel (not reticulated) veins which they contain.

24. POLYSIPHONIA BYSSOIDES, *Greville.*

Hab. On rocks, &c., in the sea. Annual? Summer and autumn. It is said to be abundant on the eastern and southern shores of England and Ireland; but rare in Scotland and in the west of Ireland. We can state that during some seasons it is very frequent on the coast of Ayrshire.

The time of the year when it is most abundantly got is in the month of August, when it is driven out by the tide in large tangled masses, often intermingled with the beau-

69.

70.

71.

72.

h bth.

B.B.&R.

tiful zoophyte, *Valkeria cuscuta.* Neither are we sure that it is annual; for we have often got it in winter in small patches of a very dark hue, and stripped of almost all its pretty byssoid ramuli. When got fresh in summer it is of a fine clear colour. It generally loses this when exposed to the air, or put into fresh water and dried. Occasionally, however, it retains its clear red colour in drying. We wish we could discover the secret for fixing this fine colour, as it adds much to the loveliness of this beautiful plant.

There is a variety at times found both in Arran and Ayrshire, in which the branches are opposite and horizontal, so that in form the specimens greatly resemble larch-trees. This variety is generally of a darker hue than the common kind.

Genus LXXXVI. DASYA, *Agardh.*

Gen. Char. Frond filamentous; the stem and branches mostly opake, irregularly cellular (rarely pellucid, longitudinally tubed), composed internally of numerous parallel tubes; the ramuli jointed, single-tubed. Fructification twofold, on distinct plants: 1, ovate capsules (ceramidia) furnished with a terminal pore, and containing a tuft of pear-shaped spores; 2, lanceolate pods (stichidia) containing tetraspores ranged in transverse bands.—The name is from a Greek word signifying *hairy.*—*Harvey.*

1. DASYA COCCINEA, *Agardh*.

Hab. In the sea, frequent. Annual. Summer and autumn. In England, Mrs. Griffiths, Professor Walker Arnott; in Ireland, Miss Hutchins; in Bute and Frith of Forth, Dr. Greville; said to be rather rare in Scotland, but we do not think so. It is not equally abundant at every season; but is generally frequent during summer and early autumn about Ardrossan, Troon, Ayr, Ballantrae, Portpatrick, in the island of Arran, island of Cumbrae, and at Southend, Kintyre.

The most beautiful specimens I ever saw were from Ballantrae. It is thought to be annual. I question this, having often got it in winter. I found it on the 22nd of last December, full size, covered with *Crisia denticulata*, and just coming into fruit, some of the red-tipped stichidia having the granules quite formed. It is one of the most beautiful of our Algæ, on account of its fine crimson colour and feathery form. It is found here occasionally in its young state, in which at one time it was called *Ceramium patens*.

There are very fine figures of it in Dillwyn's 'British Confervæ,' Pl. XXXVI. He says, "Few marine productions exceed the present species in beauty or frequent occurrence, and none meets with more general admiration, or is more frequently gathered and used in ornamental devices by the female visitors on our shores."

2. DASYA OCELLATA, *Harvey*.

Hab. On mud-covered rocks in the sea. Rare. Annual. Summer. Discovered by Mrs. Griffiths on the pier at Torquay; Professor Walker Arnott, Whitsand Bay; Trevot, Rev. Mr. Hore; in Ireland by Professor Harvey, Mr. Andrews, Mrs. Grey, Miss Gower.

"The specific name was no doubt intended by Grateloup to allude to the eye-like spots caused by the density of the ramuli at the tips of the branches. The branches resemble delicate feathers marked with an eyelet."—See Phyc. Brit., Plate XL.

3. DASYA ARBUSCULA, *Agardh*.

Hab. Rocks near low-water mark. Annual. Summer.

A beautiful little plant. Rare in England. Not rare in Ireland, and said not to be rare in the west of Scotland; but we have seen only one specimen which was obtained in Ayrshire by Miss M'Leish. See beautiful figure, Pl. CCXXXIV. 'Phycologia Britannica.'

4. DASYA VENUSTA, *Harvey*. (*spec. nov.*)

Hab. Cast on the shore. Annual. Summer and autumn

Very rare. Discovered by Miss White and Miss Turner on the shores of Jersey. We have been favoured with a specimen from Miss Turner.

This is a plant of surpassing loveliness, and a great addition to our marine flora. For an excellent description and figure I refer to Phyc. Brit., Pl. ccxxv.

Series III. CHLOROSPERMEÆ.

"It is the midnight hour:—the beauteous sea,
  Calm as the cloudless heaven, the heaven discloses,
While many a sparkling star, in quiet glee,
  Far down within the watery sky reposes.
    As if the ocean's heart were stirr'd
    With inward life,—a sound is heard,
      Like that of dreamer murmuring in his sleep:
  'Tis partly the billow and partly the air,
  That lies like a garment floating fair,
    Above the happy deep.
Oh! thou art harmless as a child,
Weary with joy, and reconciled
  For sleep to change its play;
And now that night has stayed thy race,
Smiles wander o'er thy pleased face,
  As if thy dreams were gay."

                WILSON's *Isle of Palms.*

Family XVIII. SIPHONEÆ.

Genus LXXXVII. CODIUM, *Stackhouse*.

Gen. Char. Frond green, sponge-like (globular, cylindrical or flat, simple or branched), composed of tubular, interwoven, inarticulate filaments. Fructification, opake vesicles attached to the filaments.—The name is from a Greek word signifying the *skin of an animal.—Greville.*

1. CODIUM ADHÆRENS, *Agardh*.

Hab. On marine rocks. Perennial. Summer and winter. Rare. In England, Mrs. Griffiths, Mr. Ralfs, Miss Warren, Mr. Peach; in Ireland, Mr. D. Moore, and Mr. G. Hyndman. See Phyc. Brit., Pl. xxxv. A.

2. CODIUM AMPHIBIUM, *Moore*.

Hab. On turf-banks at extreme high-water mark. Discovered by Mr. M'Calla near Roundstone Bay.

3. CODIUM BURSA, *Agardh*.

Hab. On rocks in the sea. Rare. In England, coast of Sussex, plentifully, Hollas, Turner; Cornwall, Mr. Rashleigh; Torquay, Mrs. Griffiths; Brighton, Mr. M. C. Pike; Belfast, Mr. Templeton. See Dr. Greville's excellent description, 'Algæ Britannicæ,' p. 186.

4. CODIUM TOMENTOSUM, *Stackhouse*. (Plate XVI. fig. 63.)

Hab. On rocks in the sea, and in rock-pools near high-water mark. Perennial. Said to be common on all the British shores, but we do not think that it is so on the western shores of Scotland. My friend Dr. Curdie, now in Australia, sent it to me from the island of Gigha, off Kintyre. We know that it was got by Mr. W. Thompson at Ballantrae, in Ayrshire. Dr. Greville mentions that it was found by Miss Hutchins in Iona. We procured it once in Arran, betwixt Brodick and Corrie, in a rock-pool; but these are all the Scottish habitats that we know of.

It is not beautiful, but it has an uncommon appearance, more like a sponge than an Alga. It clings so firmly to the rock that it requires some effort to detach it. I was gratified by finding *Codium tomentosum*, which I had never before seen, except in a dried state; but I was still more pleased with what I saw upon it. On taking it out of the water, I observed a greenish gelatinous animal on it, which, without examination, I cast into the pool again, that it might continue to enjoy life. I afterwards saw two more, and observing that they were beautifully mottled with azure spots, I placed them in my vasculum, along with some branches of the *Codium*, and on afterwards putting them into a tumbler of sea-water, I found that I had got a rare and lovely mollusk, discovered by Colonel Montagu on the coast of Devonshire, and described by him in the Transactions of the Linnean Society. It was *Actæon viridis*, the Green Actæon. This is not the place to describe it, but I have attempted to do so elsewhere.* See instructive figures in Alg. Brit., Pl. xix., and Phyc. Brit., Pl. xciii.

Genus LXXXVIII. BRYOPSIS, *Lamour.*

Gen. Char. Frond membranaceous, filiform, tubular, cylindrical, glistening, branched, the branches imbricated or distichous, and pinnated, filled with a fine green minute granuliferous fluid. —The name is from two Greek words signifying *the appearance of a moss.*—*Greville.*

1. BRYOPSIS PLUMOSA, *Lam.* (Plate XVI. fig. 64, frond, natural size, and a portion magnified.)

Hab. In the sea on stones and rocks, and in rock-pools. Annual. Summer. Scotland, England, and Ireland. Not very common in Scotland. I first procured it at the Black rocks, Troon, in Ayrshire. I afterwards found at Saltcoats the largest and richest specimens I have ever met with. Gathered by the Rev. G. Laing and by Mr. R. M. Stark at Joppa, near Edinburgh, and by Mrs. Stark, at Dunbar.

* 'Excursions to Arran,' by the Author.

This is one of the most attractive of our Algæ. The colour is a rich glossy green; the form is symmetrical, and resembles the feathers of a green parrot. The first notice of this plant is by Hudson, in his 'Flora Anglica.' See good figures of it in Alg. Brit., Pl. xix., and in Phyc. Brit., Pl. iii.; the latter, though beautiful, is not glossy enough, and the green is lighter than in our Ayrshire specimens.

2. BRYOPSIS HYPNOIDES, *Lam.*

Hab. In rock-pools near low-water mark, or on other Algæ in deep water. Annual. Summer. Very abundant in the west of Ireland beyond tide-mark on *Laminaria saccharina.*

This is much rarer in Scotland than *Bryopsis plumosa.* Found by Sir Wm. Jardine at Southerness, Kirkcudbright; by Dr. Hasell, Prestonpans; by Mr. R. M. Stark at North Berwick; and by Isabella Landsborough at Seamill, north of Ardrossan, in considerable abundance in rock-pools. It is more branchy, more slender, more flaccid, and of a yellower green than *B. plumosa*, and on the whole less beautiful. The specimens got at Seamill were fine large ones, and comparing them with the excellent specimens of *B. plumosa* found at Saltcoats, I had no doubt at the time that they were distinct. The colour and the branching were very different; the ramuli, instead of being distichous, as in *B. plumosa*, were irregularly scattered, and proceeded from all sides of the filaments, having, in the water, a bushy appearance, like a fox's tail. Afterwards, however, this opinion was considerably shaken, as I found the two running into each other, so that I could not distinguish them.

Genus LXXXIX.   VAUCHERIA, *De Candolle*.

Gen. Char. Fronds aggregated, tubular, continuous, capillary, coloured by an internal green pulverulent mass. Fructification, dark green, homogeneous vesicles (*coryocistæ*, Ag.) attached to the frond.—*Greville*.—The name is in honour of M. Vaucher, a distinguished naturalist, author of a standard work, '*Histoire des Conferves d'eau douce.*'

1. *Vaucheria submarina*, Berk.   3. *V. velutina*, Agardh.
2. ——— *marina*, Lyngbye.

These three *Vaucheriæ* have been got only at Appin, Argyleshire, and in the south of England. See Harvey's 'Manual of British Algæ,' page 195.

Family XIX. CONFERVEÆ.

> "L' onda dal mar divisa
> Bagna la valle e 'l monte;
> Va passagiera in fiume,
> Va prigioniera in fonte;
> Mormora sempre, e geme,
> Fin che non torna al mar:—
> Al mar, dov' ella nacque,
> Dove acquistò gli umori,
> Dove da' lunghi errori
> Spera di riposar."—METASTASIO, *Artaserse*.*

Genus XC. CLADOPHORA, *Kützing*.

Gen. Char. Filaments green, jointed, attached, uniform, branched. Fruit, aggregated granules or zoospores, contained

---

* "Waters, from the ocean borne,
 Bathe the valley and the hill;
Prisoned in the fountain, mourn,
 Warble down the winding rill;
But wherever doomed to stray,
 Still they murmur and complain,
Still pursue their lingering way,
 Till they join their native main.
After many a year of woe,
 Many a long, long wandering past;
Where at first they learned to flow,
 There they hope to rest at last."
     BEATTIE, *Translation*.

73.

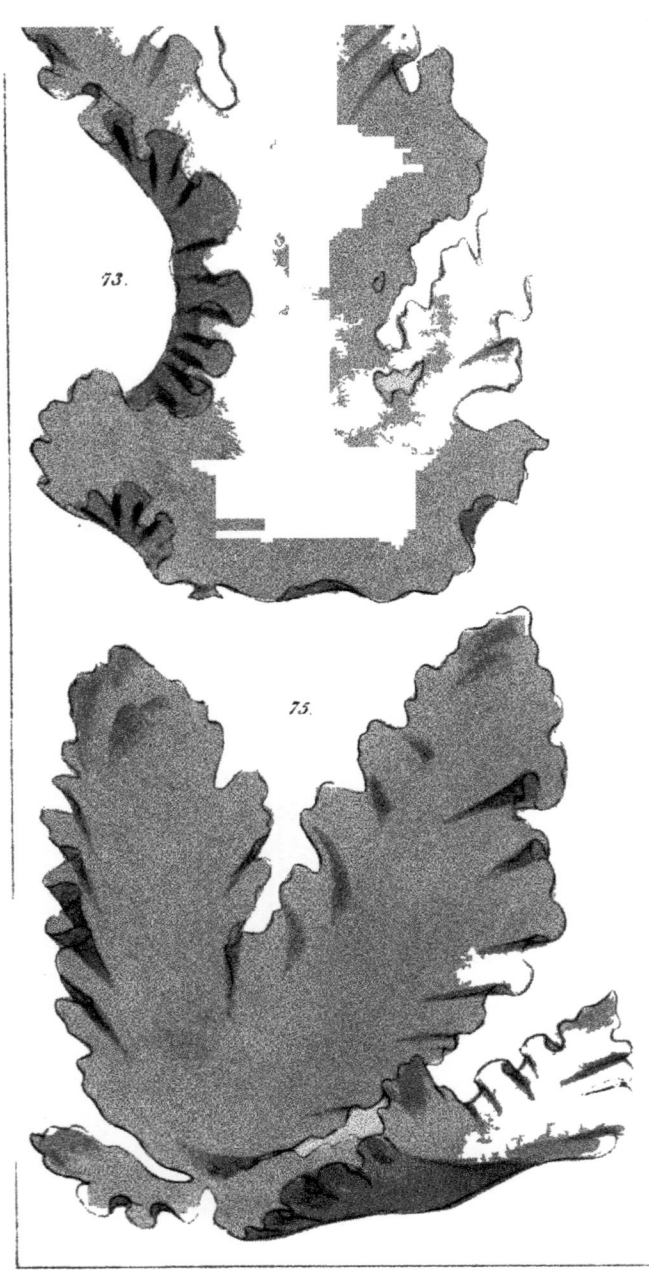

75.

Fitch lith.

77.

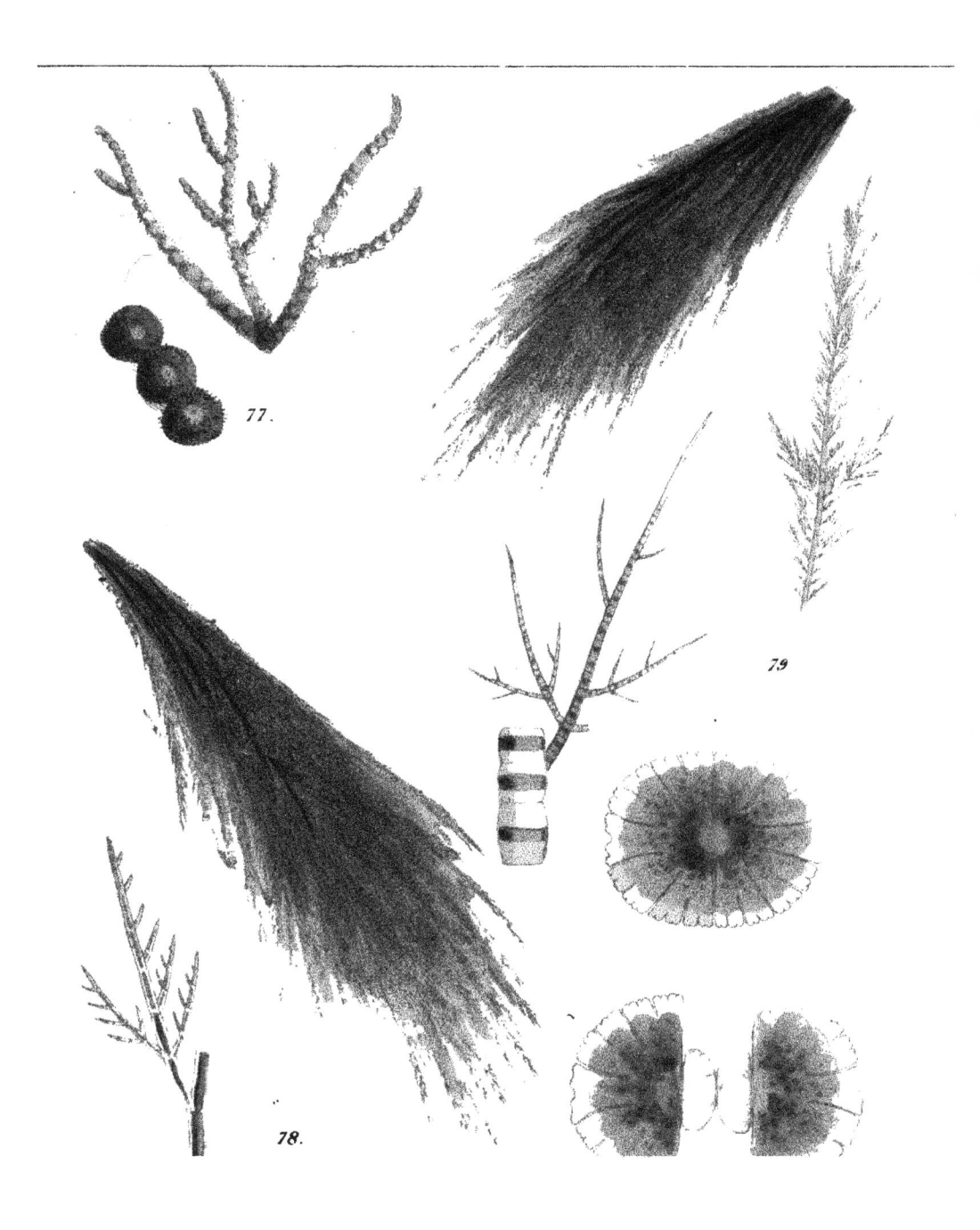

78.

79.

in the joints, having at some period a proper ciliary motion.—The name means *branch-bearing*, *Conferva* being retained for the species with simple filaments.

1. CLADOPHORA RUPESTRIS, *Kütz*. (Plate XVII. fig. 67, branches, of natural size, and a branchlet, magnified.)

Hab. On rocks in the sea, from high-water mark, and often beyond that. Annual. Summer. Very common and variable.

Near high-water mark it is a plain-looking plant, closely tufted, and of a dirty greyish-green colour. When obtained in favourable circumstances, in deep rock-pools, or by dredging from deep water, it is truly a lovely plant, of a fine dark green. The only draw-back to it is that it does not adhere well to paper in drying; and this, we doubt not, is the reason why this beautiful species is less frequently seen in collections than we might expect.

2. CLADOPHORA RECTANGULARIS, *Griffiths*. (Plate XVII. fig. 68, plant, natural size; to the right, at the bottom of the plate, there is a portion of a branch magnified.)

Hab. In the sea, in deep water. Annual. Summer. Torquay, Mr. Borrer, Mrs. Griffiths; Galway, Mr. Reilly; it has not been found in Scotland; in England it is very rare, and it was thought to be so in Ireland, till it was discovered by Mr. M'Calla in Roundstone Bay in such abundance that it was carted away for manure. At the depth of four or five fathoms it covers the bottom to a considerable extent.

It is about a foot long, and cannot be mistaken for any other species, being so easily known by its patent, opposite branches and branchlets. It is of a light green colour. It does not adhere well to paper in drying. See in 'Phycologia Britannica,' Pl. xii., a good figure of var. β, with longer branches.

3. CLADOPHORA PELLUCIDA, *Kütz.*

Hab. On the bottom and sides of deep rock-pools near low-water mark. Annual? Summer. Not uncommon on the shores of England and Ireland. We do not know that it has been found in Scotland. I have a specimen of it from Mr. William Thompson.

It is a well-marked species, as may be seen from Phyc. Brit., Pl. CLXXIV.

4. CLADOPHORA LANOSA, *Kütz.* (Plate XVII. fig. 65, two fine tufts, natural size; and on the right hand a portion of a filament, magnified.)

Hab. In the sea on rocks and on other Algæ. Annual. Summer. Common.

The best specimens for the herbarium are got on *Zostera marina*. See a fine figure of it in 'Phycologia Britannica,' Plate VI.

5. CLADOPHORA ARCTA, *Kützing.* (Plate XVII. fig. 66, tufts, of the natural size; and on the left a portion of a filament magnified.)

Hab. On rocks from half-tide level to low-water mark. It is questioned whether it is perennial; I am disposed to think that it is annual. It is very abundant in spring and summer, and the young plants, on account of their lively green, and the fine silky, silvery gloss towards their tips, are exceedingly attractive. So early as the 5th of March in 1849, I gathered beautiful tufts of it at Ardrossan. When dried in this young state it adheres well to paper, and long retains its glistering appearance at the tips; whereas, when more advanced, it loses its fine colour at the base, and soon ceases to glister in the herbarium. By the end of summer it is seldom seen, and, when found, is so coarse and faded that few would think of retaining it. *C. arcta* is now understood to include *C. Vaucheriæformis* as its young, and *C. centralis* as its old state. Substance, when young, tender and flaccid; when old it does not adhere well. See good figures of it, older and younger, in Phyc. Brit. Pl. cxxxv.

6. CLADOPHORA HUTCHINSIÆ, *Harvey.*

Hab. On rocky bottoms of clear tide-pools. Annual. Summer. Rather rare. Bantry Bay, Miss Hutchins; Larne, Dr. Drummond; Tor Abbey, Mrs. Griffiths; Belfast Bay, Mr. Wm. Thompson; Salcombe, Mr. Ralfs; Ardrossan, Major Martin; Saltcoats, Isabella Landsborough.

A beautiful species, discovered in Ireland by Miss Hutchins, whose name is held in grateful remembrance in all parts of the world. See a fine figure of this species in Phyc. Brit., Pl. CXXIV.

7. CLADOPHORA LÆTEVIRENS, *Kützing.*

Hab. On rocks and on other Algæ at low-water mark, and near high-water mark. Annual. Summer. Common.

It is frequent in the early part of the summer, on the coast of Ayrshire. Professor Harvey seems to think that it is the same as *C. glomerata,* but he yields to Mrs. Griffiths, who is of a different opinion. We have observed that *C. glomerata* retains its colour in the herbarium much better than *C. lætevirens.*

8. CLADOPHORA GRACILIS, *Griffiths.*

Hab. On rocks and Algæ. Common on the coast of Ayrshire in early summer, on Algæ; colour, a yellowish green, becoming paler in the herbarium, but retaining a silky gloss.

See a fine figure, Phyc. Brit., Plate XVIII.

9. CLADOPHORA REFRACTA, *Kützing.*

Hab. In rocky pools near low-water mark, or in deeper water. Annual. Summer. Kilkee, Professor Harvey; Ilfracombe, Mrs. Griffiths; Budleigh, Miss Cutler; Mount's Bay, Mr. Ralfs; Cork, Miss Ball; Howth, Miss Gower; Giants' Causeway, Mr. W. Thompson; D. L., Saltcoats.

See a beautiful figure of it in Phyc. Brit., Pl. XXIV. The green is darker in specimens found on the coast of Ayrshire.

Our limits compel us to give only the names of the following, for the descriptions of which I refer to Professor Harvey's 'Manual of British Algæ,' and to his 'Phycologia Britannica.'

10. *C. Brownii*, Harv.
11. — *Macallana*, Harv.
12. — *diffusa*, Kütz.
13. — *nuda*, Harv.
14. — *flexuosa*, Dillw.
15. *C. Rudolphiana*, Kutz.
16. — *albida*, Huds.
17. — *uncialis*, Harv.
18. — *glaucescens*, Griffiths.
19. — *falcata*, Harv.

Genus XCI. RHIZOCLONIUM, *Kützing*.

Gen. Char. Filaments green, jointed, uniform, decumbent, simple, or spuriously branched. Fructification, granules contained in the cells.—Name from the *root-like* form of the branches.—*Harvey*.

1. RHIZOCLONIUM RIPARIUM, *Kützing*.
Hab. Sand-covered rocks near high-water mark. Annual.

Genus XCII. CONFERVA, *Plin*.

Gen. Char. Filaments green, jointed, attached or floating, unbranched. Fructification, aggregated granules or zoospores contained in the joints, having at some period a proper ciliary motion.—*Harvey*.

1. CONFERVA TORTUOSA, *Dillwyn*. (Plate XVIII. fig. 69, plant natural size, and under it one of the filaments, magnified.)

Hab. On submarine rocks; also in salt pools. Mr. Dillwyn at Swansea; Prof. Walker Arnott, and Dr. Greville, Frith of Forth; Professor Harvey, Skerries; D.L., Saltcoats.

2. CONFERVA MELAGONIUM, *Web. and Mohr.* (Plate XVIII. fig. 70, filaments, natural size: and to the right, a portion of a filament, magnified.)

Hab. On rocks near low-water mark.

On the coast of Ayrshire it is found at mid-tide level, and it is so rigid that it stands erect when left by the tide. It is not common, though widely dispersed.

3. CONFERVA ÆREA, *Dillwyn*.

Hab. On sand-covered rocks at mid-tide level.

This resembles the preceding, but it is smaller, less rigid, and of a yellower green. They are well figured together in Phyc. Brit. Pl. xcix.

4. CONFERVA LINUM, *Roth*.

Hab. In salt-water ditches near the coast.

This species has usually been called *C. crassa*, but it is now found that *C. Linum* and *crassa* are the same. We have got it abundantly in brackish water ditches at Troon. See it and *C. sutoria* figured together in Phyc. Brit. Pl. cl.

Family XX. ULVACEÆ.

"Before me lay the sea;
Broad heaving billows murmur'd carelessly,
O'er wave-ribb'd sands, with lulling peaceful sound;
While snow-white sea-gulls sailed athwart the sky.

The air was motionless, till gentle breeze
Sprang up at sunset; yet huge lumbering waves
Rolled in from distant storm,—wild—musical!
Wave-music."—SYMINGTON's *Harebell Chimes.*

Genus XCIII. PORPHYRA, *Agardh.*

Gen. Char. Frond plane, exceedingly thin, and of a purple colour. Fructification: 1, scattered sori of oval seeds; 2, roundish granules, mostly arranged in a quaternate manner, and covering the frond.—*Greville.*

1. PORPHYRA LACINIATA, *Agardh.* (Plate XIX. fig. 75, frond, natural size.)

Hab. On rocks within tide-marks. Annual.

From spring till the end of autumn; indeed, though not in abundance, it is to be found here during all the winter. These winter specimens are very dark-coloured, and do not adhere to paper in drying. It has been stated that owing to the shrinking of the delicate fronds, this fine plant does not at any time adhere well to paper; by proper management none adheres better, and the mode of effecting it we shall state at the end of this volume, when giving directions for preserving Algæ for the herbarium. In substance it is very thin and membranaceous, and children call it *sea-silk.* It varies much in colour; the most general is the bluish-purple, in the elegant specimen in Phyc. Brit., Pl. xcii. Fine specimens of it, of this purple colour, have been gathered for us on the island of Lismore, Argyleshire, by Miss Cowan, Airds House. As got with us about Ardrossan and Saltcoats, it is dark-olive when fresh from the sea, and in drying becomes a very light olive. The most beautiful specimens I have seen of it were found by Miss M'Leish in the Clyde, at Port Glasgow; in drying they became a rich pinky red; this colour might be owing to the copious intermixture of fresh water so far from the sea.

This species, along with the succeeding, is brought to table in England and Scotland under the name of *laver*; in Ireland it is called *sloke*. We have tried it, but, though like *greens*, under the simple cooking of our *cuisinière* it had too great a smack of the sea for our taste. Under proper treatment, however, we believe that it can be rendered a grateful luxury. It requires many hours' stewing to render it sufficiently tender. Lightfoot mentions that " the inhabitants of the Western Islands gather it in the month of March, and after pounding, and stewing it with a little water, eat it with pepper, vinegar, and butter; others stew it with leeks or onions. In England it is generally pickled with salt, and preserved in jars, and when brought to table, is stewed, and eaten with oil and lemon-juice." Professor Harvey, in 'Phycologia Britannica,' says: "After many hours' boiling, the frond is reduced to a somewhat slimy pulp, of a dark brown colour, which is eaten with pepper and lemon-juice or vinegar, and has an agreeable flavour to those who have conquered the repugnance to taste it, which its great ugliness induces, and many persons are very fond of it. It might become a valuable article of diet, in the absence of other vegetables, to the crews of our whaling vessels sailing in high latitudes, where every marine rock at half-tide abundantly produces it. In its prepared state it may be preserved for an indefinite time in closed tin vessels." We regard this as a valuable hint.

2. PORPHYRA VULGARIS, *Agardh*.

Hab. On rocks and stones, between tide-marks. Common.

Except that the frond is undivided, this species does not seem to differ from *P. laciniata*. It is more elegant in form, and sometimes the colour is livelier. It approaches in shape the frond of *Laminaria phyllitis*. It is from one to two feet in length, and from two lines to two or three inches in breadth. Professor Harvey, who expressed doubts in the Manual of the distinctness of *P. linearis* as a species, gives now in the 'Phycologia' his decided opinion that it is only a narrow-fronded variety of *P. vulgaris* in a young state; that though in November they appear distinct enough, by the end of spring it will be difficult to trace in them the slender ribbons of winter. He admits that there are localities near high-water mark where the frond never attains any great length or breadth, and therefore remains more true to the name *linearis*, but the stunted growth is clearly referable to deficient nourishment. All this may be correct, but I am not thoroughly convinced; and I am unwilling to give up the pretty little *linearis*. It appears to me more distinct as a species than *P. laciniata*; and this, perhaps, is not saying much, for with us *P. vulgaris* and *P. laciniata* run very close into each other. With us they are of the same colour, and this cannot be said of *P. linearis*, which is always different from both the others; these two are never reddish here, and *linearis* always is. *P. linearis*, for many miles on the coast of Ayrshire, is got only on one little patch of rock in early spring, painting it red when the tide is out; *P. vulgaris* is found at the same season quite near it, as near high-water mark, almost as narrow as *linearis*, but nearly a foot long, which is more than the length of *linearis*; and while *linearis* is always red, *vulgaris* at that season is a pale yellow, the one continuing red and the other yellow, when dried. *P. linearis* may be got in abundance during the summer, becoming light olive when fully grown. By the middle of April every frond of *linearis* has vanished, and not one of the little pink sparklers is seen on the rocks till the succeeding spring. There are good figures of *P. vulgaris* and *P. linearis* in 'Phycologia Britannica,' Pl. ccxi. With us, however, they are both much more glossy; and *P. vulgaris*, especially early in the season, is as light, and even more yellowish, than *Laminaria phyllitis*. I see that Dr. Greville, in Alg. Brit., says that the different arrangement of the granules precludes *P. linearis* from being regarded as a miniature of either of the two.

3. PORPHYRA MINIATA. We have a strong suspicion that this is not an Alga at all. We got at Ardrossan a beautiful specimen that seemed to answer the description given by Capt. Carmichael. We spread it on paper and it seemed beautiful, but, from the fishy flavour it sent forth, we *guess* that it was neither more nor less than the floating membrane of a *Medusa*, and a cat that stole it seemed to be of the same opinion.

Genus XCIV. BANGIA, *Lyngbye.*

Gen. Char. Frond flat, capillary, membranaceous, of a green, reddish, or purple colour. Fructification, granules arranged more or less in a transverse manner.—Named *Bangia*, in honour of Hoffman Bang.—*Greville.*

1. BANGIA FUSCO-PURPUREA, *Lyngb.*

Hab. On rocks and planks of wood in the sea. Annual. Spring and summer. Brighton, Mr. Borrer; Cornwall, Mr. Rashleigh; Bantry Bay, Miss Hutchins; Frith of Forth, Professor Walker Arnott; Isle of Bute, Dr. Greville; Isle of Arran, D. L., jun.; Portincross, Ayrshire, Major Martin.

Fronds growing in a tufted manner, and generally covering a considerable extent of rock; colour, a fine dark purple, which it retains in drying. A curious circumstance is that it is found in beauty both in the sea and in fresh-water rivers.

It is generally thought that this plant is rare in Scotland. We were once of that opinion, but have since had reason to think that we had been treading upon it every spring without observing it. When the tide is out, it clings so closely to the rocks that it may be passed without notice. Now that it has been detected, it may be seen in April and May at Saltcoats in great abundance on flattish rocks, and also on the face of perpendicular rocks near high-water mark. I first procured it at Saltcoats in a rock-pool near the mouth of the harbour; Major Martin finds it of a very dark colour at Portincross.

2. BANGIA ? CERAMICOLA, *Lyngb.*

Hab. In rocky pools on small Algæ. Appin, Capt. Carmichael. We have found this at Portincross, and in Arran.

The finest specimens have been found at Saltcoats by Miss M'Leish, on *Rhodomela subfusca.* The filaments are very fine and of a beautiful rose-colour. It appears in August, and is very evanescent, for in a few days it loses its fine colour, and soon disappears.

Genus XCV. ENTEROMORPHA, *Link.*

Gen. Char. Frond tubular, hollow, membranaceous, of a green colour and reticulated structure. Fructification, three or four roundish granules aggregated in the reticulations.—The name signifies *in the form of an entrail.*

1. ENTEROMORPHA INTESTINALIS, *Link.* (Plate XVIII. fig. 72.)

Hab. In the sea, and also in brackish and fresh-water ditches. Annual. Summer. Common.

It varies in length and breadth; sometimes short and narrow, at other times two feet long and three inches in diameter. It is always simple, whereas *E. compressa* is branched. It changes from pale green to yellow, and becomes white in decay.

The most interesting specimens I ever got of this plant might be regarded as subfossil. They were found in a sandstone quarry at Ardeer, Ayrshire, about a mile from the sea. When the quarriers had removed about six feet of earth, they came to a thick stratum of shale, which was perforated in many places, and in the mouth of the bore, which was about an inch in diameter, I generally found a pretty entire specimen of *Pholas crispata*, the very same kind of mollusk that I had often seen boring the shale in the sea at Saltcoats. The perforation was about six inches deep, and at the bottom of the bore there was in all cases a matted pellet of vegetable substance. On macerating this, it spread, and I saw by its puckerings and reticulations that it was *Enteromorpha intestinalis*. I sent it to Sir William Hooker, and he said that it was correctly named. See a fine figure of this plant in 'Phycologia Britannica,' Pl. CLIV.

2. ENTEROMORPHA COMPRESSA, *Greville*. (Plate XVI. fig. 62.)

Hab. Rocks in the sea. Annual. Spring and summer. Very common.

Very variable in length and breadth. By comparing the fine figure of the preceding in 'Phycologia' with an excellent figure of this species by Dr. Greville in Alg. Brit., Pl. xix., or our own, it will be seen that there is cousiderable similarity; both are green, but this one is a darker green; both are attenuated towards the base, and both are rounded at the top; but the former is always simple, and this one is more or less branched. In drying, it does not adhere well to paper. When it decays it becomes purely white, and in this state children call it *sea-thread*.

2. ENTEROMORPHA ERECTA, *Hooker*. (Plate XVIII. fig. 71, plant, natural size, with a branch on the left, magnified.)

Hab. In the sea, about half-tide level. Annual. Spring and summer. Appin, Capt. Carmichael; Bute and Frith of Forth, Dr. Greville; Torquay, Mrs. Griffiths.

This is one of the finest of the *Enteromorphæ*, when from deep water. There are several other *Enteromorphæ*, but the most distinguished botanists are of opinion that they are varieties of one species. "Few plants," says Professor Harvey, in his Manual, "are so sportive in size and ramification, and if all the varieties were described, the species might easily be multiplied till we should have one for almost every marine pool!"

Genus XCVI. ULVA, *Linnæus*.

Gen. Char. Frond membranaceous, of a green colour (in some cases saccate, and inflated in the young state). Fructification, minute granules mostly arranged in fours.—*Greville*. The name is from the Celtic word, *ul, water*.

1. ULVA LATISSIMA, *Linn*. (Plate XIX. fig. 73.)

Hab. On rocks and stones in the sea. Annual. To be found, however, more or less during all the winter.

Fronds from four to eighteen inches long, and of considerable breadth, of an oblong shape, and waved at the edges. It is tender in substance, of a deep green colour. It adheres well to paper in drying in the early part of the season, the shading produced by the waved margin adding to the beauty of the specimens. It is called *oyster-green*, because employed to cover oysters. It is also called *green laver*, as it is employed for culinary purposes when *Porphyra* cannot be procured, but it is not thought to be so good as that. I learn from Mr. Fleming, Rector of the Academy at Kirkcudbright, that both *Porphyra* and *Ulva latissima* grow in the Dee there so far up the river that, according to the tide, they are alternately inhabitants of sea-water and fresh-water.

2. ULVA LACTUCA, *Linnæus*.

Hab. On rocks, &c., in the sea. Annual. Spring and summer. Pretty common. Devonshire; Bute; Ayrshire.

It first it is saccate, but soon bursts and becomes cleft. It is much tenderer than the former, of a lighter green, and a little gelatinous, so that it adheres well to paper in drying. Dr. Greville, in Alg. Brit., says, "It adheres so closely to paper as to resemble a drawing, and the surface shines as if varnished. When preserved in the herbarium, it is infinitely more beautiful than *Ulva latissima*." This is not according to our experience in the west; for in my opinion *U. latissima* forms the handsomest specimen, being of a richer, darker green, and more glossy than *U. Lactuca*, adhering nearly as well when young and recent. See 'Phycologia Britannica,' Pl. CCXLIII.

3. ULVA LINZA, *Linnæus*.

Ulvaceæ.] CALOTHRIX.

Hab. Rocks and stones in the sea, between tide-marks. Annual. Summer and autumn.

This is a most beautiful and truly elegant plant. The fronds are linear-lanceolate, attenuated at each extremity, and curled and waved at the margin. It is very graceful in its native element, and scarcely less so in the herbarium, when well prepared. The shadings produced by the plaiting of the curled margin in drying add much to the beauty of the plant. In early summer it is of a rich lively green, with a fine gloss, adhering well to paper. In the end of summer the green has become darker, and it is still glossy, but does not adhere well to paper. The substance is thin, and composed of two closely-united membranes. It is at times eighteen inches in length, and an inch in breadth. It grows in clusters of more than a dozen fronds. See 'Phycologia Britannica,' Pl. xxxix.

Without following further Professor Harvey's systematically arranged List, I shall in the close mention a few plants that are not included in his catalogue.

Genus XCVII. CALOTHRIX, *Agardh*.

Gen. Char. Filaments destitute of a mucous layer, erect, tufted or fasciculate, fixed at the base, somewhat rigid, without oscillation. Tube continuous; endochrome green, densely annulated, at length dissolved into lenticular sporidia.—The name signifying *beautiful hair*, the filaments being very slender and delicate.—*Harvey*.

1. CALOTHRIX CONFERVICOLA, *Agardh*.

Hab. On marine Algæ.

This is very common on sea-weeds of the smaller kinds, giving them a bluish-green hue, only a line or two high, and forming scattered or continuous tufts. There are upwards of a dozen species of *Calothrix*, some of which are rare, and some are found in fresh water. See Phyc. Brit., Pl. LVIII., LXXVI.

Genus XCVIII. RIVULARIA, *Roth*.

Gen. Char. Frond globose or lobed, rarely increasing, fleshy or gelatinous, firm, composed of continuous inarticulate filaments, annulated within, and surrounded by, or set in, gelatine.—*Harvey*.—The name is in allusion to the fresh-water habits of many of the species. Many, however, are found in the sea.

1. RIVULARIA ATRA, *Roth*.

Hab. On rocks, stones, corallines, and Algæ, between tide-marks. Perennial. Very common.

It is probable that many have observed it without supposing that it was a plant. In some localities a person can scarcely take up a handful of *Corallina officinalis* without having on it more than one specimen of *R. atra*, like little hemispherical masses of dark-green jelly. By the application of a powerful lens it may be seen that these little green pea-like masses are organic, and are made up of filaments radiating from the centre. Each one, when greatly magnified, is not unlike a handsome switch, the upper half of which is nicely knotted. See Phyc. Brit., Pl. ccxxxix.

Genus XCIX. LYNGBYA, *Agardh*.

Gen. Char. Filaments destitute of a mucous layer, free, flexible, elongated, continuous, decumbent. Endochrome (green or purple) densely annulated, and finely separating into lenticular sporidia.—Name in honour of H. C. Lyngbye, author of an excellent work on the Algæ of Denmark.—*Harvey*.

1. LYNGBYA CARMICHAELII, *Harvey*.

Hab. On marine rocks, and on *Fuci*. Appin, Captain Carmichael; Torbay, Mrs. Wyatt; D. L., Saltcoats.

It is very common here. As early as the month of April

it covers the rocks, almost at high-water mark, with a dense fleece of dark-green filaments, lying flat during ebb, but waving gracefully in the water when the tide returns. See Phyc. Brit., Pl. CLXXXVI. A.

Genus C. STRIATELLA, *Agardh*.

Gen. Char. Stipes very short, filaments curved, pellucid at the articulations, striated transversely.—It takes its name from the *striated joints*.

1. STRIATELLA ARCUATA, *Agardh*.

Hab. On marine filiform Algæ; on the coast of Ayrshire this is very common, especially in spring, almost covering the plant on which it grows, and often giving it a greenish tint. When dry, this parasitic coat has a glistering metallic cast.

Genus CI. ISTHMIA, *Agardh*.

1. ISTHMIA OBLIQUA, *Agardh*.

Hab. On filiform Algæ. Frequent on the shores of England and Ireland; we know not that it has been found in Scotland. We have numerous fine specimens of it from Mrs. Lyon, gathered by her on the shore at Glenarm, where it seems very abundant, rendering *Polysiphonia nigrescens* quite hoary. The filaments are composed of oblique-angled frustules, curling up in a curious manner; and when seen through a lens, each portion is elegantly reticulated.

Genus CII. DIATOMA.

Gen. Char. Filaments composed of rectangular frustules, cohering at the angles, and finally separating.

1. DIATOMA MARINUM, *Lyngbye.*

Hab. Parasitic on marine Algæ in spring. Greenish when fresh; powdery and pale when dry. Common in most places; but rare on the coast of Ayrshire, where its place is occupied by *Striatella.*

Genus CIII. EXILARIA, *Greville.*

Gen. Char. Frustules rectilinear, fasciculate, or spreading in fan-shaped series, fixed at the base to a receptacle or stipes.— *Greville.*

1. EXILARIA FULGENS, *Greville.*

Hab. Parasitic on marine filiform Algæ. Devonshire, Mrs. Griffiths; Appin, Captain Carmichael; Arran and Ayrshire, D. L.; Leith, D. L., jun.

Frustules of a pale yellow lustre when recent, with a glossy lustre when dry. They radiate from a minute base, in a fan-shaped manner.

Genus CIV. LICMOPHORA, *Greville*.

Gen. Char. Frustules wedge-shaped, united into fan-shaped laminæ, fixed to the summit of a (usually branched) stipe.— The name is from two Greek words signifying *fan-bearer*, highly expressive of the form of these minute but beautiful objects.— *Hooker*.

1. LICMOPHORA FLABELLATA, *Greville*.

Hab. On marine Algæ. Bantry Bay, Miss Hutchins; Appin, Captain Carmichael; Devonshire, Mrs. Griffiths; Antrim, Mr. D. Moore; Strangford Lough, Mr. W. Thompson; Portincross, Ayrshire, D. L.

Tufts half an inch high, deep green when recent (with us very light greyish-green); when dry, grey and glistering. Dr. Greville's figure of this interesting plant is thought remarkably good.

2. LICMOPHORA SPLENDIDA, *Greville*.

Hab. Parasitic on marine Algæ, and on *Zostera marina*. Appin, Captain Carmichael.

" A very fine species, nearly allied to *L. flabellata*, but smaller, less divided, and the frustules more broadly wedge-shaped. The tufts are two or three lines in height, and often invest the whole surface of the plant on which it grows."—*Grev.* This plant had not been found by any since its discovery at Appin by Capt. Carmichael, till it was got in considerable abundance by D. L., jun., in September 1848, at low-water mark, in a little creek formed by trap dykes, in the parish of Ardrossan. When he brought it to me, I was much struck with its beauty. Hoping that it was *L. splendida*, I sent it to Dr. Greville, and was gratified by his pronouncing it to be that rare plant. Though minute, it is well deserving of the name of splendid; it is like an assemblage of hundreds of beautiful little fans. Had I believed in the existence of fairies as firmly as I did in my childish years, I could have imagined that some marine Queen Mab, and all the ladies of her court, were congregated amidst the branchlets and filaments of the little Alga. "*Materiem superabat opus :*" every fan was of exquisite workmanship. Raised on a little stem, they were spread out so as to form in some cases more than a semicircle, the rays numbering from ten to twenty-six. Each ray or frustule was wedge-shaped, and a little denticulated at the top; the upper part was amber-coloured, and as each ray had a lighter-coloured dot in the middle of this portion, these bright dots formed a crescent of sea-gems, adorning the fan. Under this amber-coloured portion there was a pellucid band, the lower part of the fan being amber-coloured, like the upper. Aided by a microscope, the whole was so beautiful that a lady to whom I showed a portion of *Licmophora* thus magnified, said she could not fall asleep for a long time that night, as the lovely fans seemed ever before her eyes; and when she did sleep she dreamed of them. What adds to the wonders of these *Diatomaceæ* is, that they are partly formed of flint, which they extract from the waters, so that, though seemingly frail, they are imperishable!

Genus CV. SCHIZONEMA, *Agardh*.

Gen. Char. Frustules in longitudinal series or scattered, and enclosed within a simple or branched, gelatinous or membranaceous frond, composed of one or several tubes.—The name is from two Greek words signifying *to divide*, and *a thread*, as the typical species are formed, as it were, by dividing the frond.—*Harvey.*

1. SCHIZONEMA OBTUSUM, *Greville*.

Hab. Parasitic on small Algæ in the sea. Frith of Forth, Dr. Greville; Appin, Captain Carmichael; Torquay, Mrs. Griffiths; coast of Antrim, Mr. D. Moore; Leith, D. L., jun.

2. SCHIZONEMA HELMINTHOSUM, *Chauv.*

Hab. On rocks in the sea. Frith of Forth, Dr. Greville; Torquay, Mrs. Griffiths; Saltcoats, D. L.

3. SCHIZONEMA IMPLICATUM.

Found at Ardrossan by Major Martin and D. L.

Mrs. Griffiths has kindly sent us above a dozen named species of this troublesome tribe.

FRESH-WATER ALGÆ.

"Let us, then, consider the works of God, and observe the operations of his hands; let us take notice of, and admire, his infinite wisdom and goodness in the formation of them: no creature in this sublunary world is capable of so doing besides man, and yet we are deficient herein. We content ourselves with the knowledge of the tongues, or a little skill in philosophy, or history, perhaps, and antiquity, and neglect that which seems to me more material,—I mean Natural History, and the works of the Creator. I do not discommend or derogate from those other studies; I should betray mine own ignorance and weakness should I do so: I only wish that they might not altogether jostle out and exclude this. I wish that this might be brought in fashion among us; I wish men would be so equal and civil as not to disparage, deride, and vilify those studies which themselves skill not of, or are not conversant in. No knowledge can be more pleasant than this,—none that doth so satisfy and feed the soul, in comparison whereto that of words and phrases seems to me insipid and jejune."—*Ray*.

Though the limited size of this work will scarcely permit us to enter on this department, we cannot think of finishing the volume without describing a few of the fresh-water Algæ.

Many may wish to learn something of Algology who are at a distance from the sea, and have no opportunity of collecting sea-plants in a recent state. It is well to let such know that fresh-water Algology opens up for them a wide field, which of late has been very successfully cultivated by British botanists. In 1845 we were favoured by Mr. Hassall with an excellent work on the History of British Fresh-water Algæ, in two volumes, one of letter-press and the other of plates; and in 1848 we have had another admirable work by Mr. Ralfs, on British Desmidieæ, in one handsome volume, containing accurate descriptions and exquisite illustrations, in thirty-five plates. To these two works we refer those who devote themselves to this study; but as many may not be disposed to purchase costly works, however excellent, till they know something of what they contain,—without attempting to treat of fresh-water plants at all systematically, we shall describe a few, as a kind of sample of the rich stores within the reach of the naturalist, however remote from the sea.

One reason, no doubt, why these plants are not more generally studied, is, that they are so minute that their beauty cannot be seen by the unaided eye. Even in the case of those whose filaments may be some feet in length, and which cannot fail to be seen, as they grow in masses of considerable breadth, the filaments are so densely crowded, that instead of being regarded as plants, they are looked upon as some green impurity, which in Scotland goes under the general name of *slaak*. When a small portion, however, of this despised slaak is taken, and laid on talc, and examined by the aid of a microscope, or even a hand lens, the person who thus beholds it will be filled with astonishment; he will see that what he regarded as shapeless filth, is of exquisite workmanship, and worthy of the Hand by which it was made; and he may learn that what he thought worse than useless, instead of polluting the waters, is one great cause of their purity and wholesomeness; that without these Algæ the waters would soon become so putrid and poisonous as to spread malaria over wide districts of country, and lay them desolate.

But these are not the only purposes they serve. They afford shelter to countless myriads of living creatures, especially infusorial animalcules, which not only enjoy life, but —minute though they be, and unseen by man,—perform wonderful functions for his benefit. They are found most abundantly in all stagnant waters:

"Where the pool
Stands mantled o'er with green, invisible
Amid the floating verdure, millions stray."

By placing a drop of water in the field of the microscope, and observing the merry evolutions of its multitudinous inhabitants, we see that He who made them blessed them with happiness. It has been long known, also, that by subsisting on the dead bodies of larger aquatic animals, they "thus limit," according to Liebig, "to the shortest possible period, the deleterious influences which the products of dissolution and decay exercise upon the life of the higher classes of animals." "The recent discoveries," he adds, "which have been made respecting these creatures, are so extraordinary and so admirable, that they deserve to be made universally known." The most remarkable fact in these discoveries is that the functions of animal life are *reversed* in these animalcules, and that, instead of evolving carbonic acid gas, as, by breathing, other animals do, they evolve pure oxygen.* The air-bubbles given out by water in which these animalcules abound, contain such pure oxygen that a small bit of deal match-wood in which a flame

* See 'Blights of the Wheat and their Remedies,' published by the Religious Tract Society.

has just been extinguished, will burst into a flame again on being immersed in any one of them. "I myself," says the distinguished German chemist, "took an opportunity of verifying this remarkable fact upon finding in a trough of water in my garden the fluid coloured green by the presence of various species of Infusoria. I filtered it through a very fine sieve, in order to separate all *Confervæ*, or vegetable matters; and then exposed it to the light of the sun in an inverted broken glass, completely full, the aperture of which was confined by water. After the lapse of a fortnight, more than thirty cubic inches of gas had collected in the glass, which proved to be very rich oxygen."

In the most extensively diffused animalcules, then, namely the green and red Infusoria, we recognize a most admirable cause, which removes from water all substances injurious to the life of the higher classes of animals; and creates in their place nutritive matters for the sustenance of plants, and the oxygen indispensable to the respiration of animals. We see, then, that the benignant purposes of God towards man are answered not only by the fresh-water Algæ, but, what is more extraordinary and had never been suspected till discovered by science, that the functions of animal nature are reversed in the millions of millions of unseen creatures that dwell among these weeds, that they may all be made subservient to the well-being of man.

After this digression respecting these animalcules, which I have often admired without thinking that in their sportive movements they were contributing to the happiness of the human race, we shall turn our attention for a little to some of the fresh-water Algæ, amidst which they are so plentifully found. We shall select some that are by no means rare, and whose beauty is very evident, even to the naked eye. We begin with one which may be found in almost every stream in all places of our land.

CLADOPHORA, *Linnæus*.

Gen. Char. Filaments rising from a scutate root, finely tufted, bushy, somewhat rigid, bright green, branches crowded, irregular, erect; the ultimate ramuli secund, subfasciculate; articulations 4–8 lines longer than broad.—*Harvey's Manual.*

1. CLADOPHORA GLOMERATA, *Linnæus*. (Plate XX. fig. 78, portion of the frond, natural size; and on the left a branchlet, magnified.)

Hab. In clear streams, wells, &c. It is very common, attaching itself to stones and sticks in streams and pure ponds.

It is a remarkably variable plant. I remember finding a beautiful variety of it in the pool of a little cascade at King's Cove, Arran. The filaments were simple, not tufted, of a fine delicate texture, and having, when dried, a soft, silky, glossy appearance, such as *Cladophora gracilis* often has. The most beautiful specimens of it I ever saw were found by D. L., jun., at Corriegills in Arran, in the month of September. They were quite of the normal type, beautifully tufted, of a lively green, and retaining all their beauty when dried. The time for getting it in the greatest beauty is in early summer, when it is in a young state, or in the autumn, when it sometimes assumes a fresh dress after the scorching heat of summer.

Mr. Hassall, in his 'British Fresh-water Algæ,' says:—" Notwithstanding that its usual resort is the stream and the waterfall, it will flourish and increase in size amazingly for weeks and months in a vessel, the water of which is occasionally renewed. I have thus kept it for many weeks, removing (when by its growth it had filled the vesssel) all but a small portion of it; this, however, speedily increased, and again filled its dwelling-place. The tearing away of portions of the plant in no way impaired the vitality of the remainder, as from its aggregation of minute cells, each the analogue of the other, might, à *priori,* have been conjectured. After the species has been confined for some time, if it be examined with a glass, very many of the filaments will be found to be increased with numerous smaller filaments. These are the young of the plant derived from the growth of zoospores which have attached themselves to the parent filaments. It was the occurrence of a specimen thus infested, that induced Vaucher to place this species in his genus *Prolifera."* It is often covered with a parasite, *Diatoma vulgare,* giving it a rich brown instead of its natural rich green colour.

VAUCHERIA, *De Candolle.*

Gen. Char. Fronds aggregated, tubular, continuous, capillary, coloured by an internal green pulverulent mass. Fructification, dark-green homogeneous vesicles (*coniocystæ,* Ag.), attached to the frond.—The name is in honour of M. Vaucher, a distinguished writer on Fresh-water *Confervæ.—Greville.*

This is a very natural and well-defined genus of plants. The structure of the frond is like that of *Bryopsis* and *Codium,* but there is no appearance of reticulation or cells. It resembles the *Confervæ* in general appearance, and it is found along with them in ditches and little waterfalls, and on damp ground. Common as *Vaucheriæ* are, they are very remarkable plants, and the investigation of them afforded great delight to the intelligent mind of Vaucher. Their power of resisting cold, and of sustaining high degrees of temperature, is very extraordinary. M. Vaucher mentions, that when he was making his experiments at Geneva, an intense frost set in, and froze the water in a vase in which his *Vaucheriæ* were kept. The frost continuing for a fortnight, he feared that, as they were enveloped in ice all that time, they would be completely destroyed; but when thaw came, he found, to his great delight, that they had sustained no injury; and he had the satisfaction of seeing the grains germinating, as if they had never known frost. Were not this power granted to them, and especially to the seeds, they would soon be exterminated, as every winter they are frozen for weeks together. Their power of withstanding great heat is scarcely less remarkable, and not less necessary for the continuance of their existence. A very compact capsule envelopes the spore, and preserves the internal moisture from being dried up. Their seed in general ripens before the drought of summer, and when the shallow pools are dried up, the seed lies in the mud, till it is called into life by the returning heat and moisture of early spring.

What is recorded by M. Unger respecting *Vaucheria clavata,* is exceedingly interesting. He set himself uninterruptedly to observe one of the tubercles of fructification, and when he had done so for half an hour, it became darker in its colour, and a little transparent at its extremity; in the middle it was somewhat contracted, and had some traces of spontaneous motion. He could scarcely believe his eyes when he perceived the contraction to become more decided, and a cavity to be formed at the base. The contraction at length divided the globule into two smaller globules, which moved spontaneously towards the summit. As the development proceeded, the cavity and the uppermost globule became enlarged, while the inferior globule became diminished; the latter at length disappeared, and the remaining large globule escaped by a terminal orifice ascending till it reached the surface of the water. The whole process occupied about a minute. On various other occasions he observed numbers of these globules swimming freely about here and there, stopping, and again setting themselves in motion, exactly like animated beings; and he does not scruple to call them infusory animalcules. We have no doubt that he is correct in his statement as to the motion and subsequent germination of the green globules; but motion such as this, however wonderful, does not prove that they had animal life. He found that the motion of the spores was effected by their surface being covered with vibratile ciliary organs; but the difficulty still remains— What moves these cilia? Cilia are not peculiar to living creatures. Verily there are mysteries in nature which philosophy cannot explain;—depths in the organism of a common plant which human intellect cannot fathom!

1. VAUCHERIA DICHOTOMA, *Agardh*.

Hab. Ditches and pools.

The frond of this is dichotomous, as the specific name implies; it is branched, about a foot in length; often filling pools with a close matted stratum. The vesicles are solitary globules, and sessile; the colour is green, lighter or darker.

There are a considerable number of *Vaucheriæ* which I must pass over. I found one lately in circumstances which seemed uncommon, and I have not yet been able to ascertain which one it is. It was growing in full fruit in November on the steep inside wall of a lime-kiln, uncovered above, along with *Funaria hygrometrica*. It was of a very dark green, the vesicles were abundant, and the filaments fine.

BATRACHOSPERMUM, *Bory*.

Gen. Char. Filaments invested with gelatine, moniliform, branched. Fruit, globules of dense filaments scattered throughout in whorls, and to which they are attached by a single filament.—The name is composed of two Greek words, signifying *frog-spawn*.

1. BATRACHOSPERMUM MONILIFORME, *Bory*. (Plate XX. fig. 77, filaments of *B. moniliforme,* and on the left a portion of moniliform filament, magnified.)

Hab. In pure water, in wells and fountains and gently flowing streams. Not very common. Mr. Ralfs, Devonshire; Mr. Hassall, Cheshunt; Dr. Dickie, Aberdeen; Dr. Greville, near Edinburgh; Mr. Keddie, Dunollybeg, near Oban; D. L., Ballantrae.

The specimens I got at Ballantrae were as fine as any I had ever seen. They filled a little fountain of water on a hill-side near the sea-shore. When brought out in handfulls from the little spring well, they were truly loathsome, or at least they would have been so to a person unacquainted with them, for they greatly resembled frog-spawn.

I knew well, however, what a prize I had got, and with the fine specimens they formed, many friends were supplied. The filaments were about six inches in length, and the specimens shaded with tints of various colours. When spread on paper, the beautiful beading of the filaments can be seen by the naked eye, but it appears still more exquisitely beautiful when a lens is applied. They are so gelatinous that in general they must be allowed to dry on the paper before any pressure is applied. Early in April this Batrachosperm makes its appearance as a light green down on stones, or sometimes on grass, floating on the edge of the pool. At a more advanced period it becomes detached, and continues for a time to grow in a free state.

There must be something peculiar in the water in which it grows, for year after year it continues to be found in the same little well, though not got in similar-looking wells for many miles around. I have tried to transplant it into other pools, but without success.

2. BATRACHOSPERMUM ATRUM, *Bory*.

Hab. In wells and little pools.

A few years ago this was found for the first time that it had been got in Scotland, by D. L., jun., but it has been found by him since in greater abundance and beauty in the parish of Stevenston, at Ashgrove Loch; and it has been lately got in a well near Beith by Mr. Levack. It has been found in several places both in the east and west of Scotland by Mr. Wyville Thomson, with filaments in some cases six or eight inches in length. It differs from *B. moniliforme* by being devoid of moniliform whorls, the distant whorls being like the commencement of filaments.

3. BATRACHOSPERMUM STAGNALE, *Hass.* This is of a pale yellowish-green colour, the branches are sometimes thick and a little compressed, and at other times slender and round. It is much branched and less gelatinous than *B. moniliforme.* It has been found by Mrs. Dobie, of Crummock, in the Saint's well, near Beith; by Mr. R. M. Stark near Edinburgh, and by D. L., jun., at Ashgrove Loch.

4. BATRACHOSPERMUM VAGUM, *Ag.* This is probably the most beautiful of all the Batrachosperms, at least in some of its states and forms. The colour is frequently of a very agreeable bluish or glaucous green. It has been got on the summit of Snowdon by Sir William J. Hooker; at Loch Phadrick, in Aberdeenshire, by Professor Dickie, 2199 feet above the level of the sea; by Captain Carmichael at Appin; by Professor W. H. Harvey in Galway, and we

have beautiful specimens of it from our excellent friend Professor Scouler, of Dublin.

OSCILLATORIA, *Vaucher*.

Gen. Char. Filaments simple, even, clustered closely, striated, and generally lying in a mucous matrix.—The name is from a Latin word signifying *to oscillate* like the pendulum of a clock, from the motion that the filaments are thought to make.

1. OSCILLATORIA LIMOSA, *Vaucher*.

Hab. Ditches and sewers by road-sides. Common.

Stratum rich dark green, very thin, gelatinous, with short rays; filaments pale green, straight; striæ rather distinct, evident.

It is not easy to determine the species in this genus; even in this one there is uncertainty. It is *O. tenuis* of Hassall, and *O. viridis* of Johnston. The family, however, of the *Oscillatorieæ* is one of the most distinct and remarkable of the divisions of the Algæ. They are distinguished by the rapidity of their growth, the brilliancy of their colours, and the peculiar motion or oscillation of their filaments, on which their generic name is founded. Mr. Hassall does not think that there is anything very remarkable in this motion, which he considers as partly external, and altogether physical. Their filaments, he says, are very straight and elastic, and when they are placed for observation on the field of a microscope, they are bent out of their natural straight line, and make an effort to recover it. Currents almost imperceptible in the liquid in which they are immersed, and perhaps unequal attractions among the filaments themselves, are causes amply sufficient to explain any motion, he thinks, that he has ever witnessed amongst the *Oscillatorieæ*. Captain Carmichael, however, a very accurate observer, is of a different opinion, though he probably makes too much of their motion when he considers it a proof of animal life. "Let a small portion of the stratum be placed in a watch-glass nearly filled with water, and covered with a circular film of talc, so that its edge may touch the glass, the *water* will be rendered as fixed as if it were a piece of ice. The glass may now be placed under the microscope, and the oscillation of the filaments viewed without the risk of disturbance from the agitation of the water; by following this course it will be speedily perceived that the motion in question is entirely independent of that cause."

2. OSCILLATORIA THERMALIS, *Hassall*.

This resembles the preceding, only that it is finer in every respect. It cannot be called common, for I know no other habitat than that in which I found it,—in a current of tepid water, flowing from a boiler at the Turf Dyke coal-pit, Stevenston. When taken out of the water it is like green jelly. When a small portion of it is placed on paper, and the paper is submerged, it almost immediately begins to put forth its bright green glossy filaments, which in the course of a few hours extend in all directions an inch or an inch and a half in length. It makes a very beautiful specimen, especially if the central patch from which the filaments spring, is removed, and the space becomes filled with filaments from the first growth. If the patch is not removed, it is, when dried, apt to crack and fall off, leaving an empty space.

3. OSCILLATORIA MUCOSA, *Bory*.

Stratum gelatinous, dark, æruginous green, glossy; filaments large; striæ subdistant.

This, like the former, seems to be a new species, as Mr. Hassall had never seen any specimens but those I sent him. I found them floating in a pool at the same coal-pit, but the water was not warm, and not very pure. They formed little filmy clouds, which were almost imperceptible.

DRAPARNALDIA, *Bory.*

Gen. Char. Filaments free, not immersed in gelatinous fluid.—*Hassall.* It is affectionately dedicated by M. Bory to his departed friend M. Draparnaud, a distinguished French naturalist.

1. Draparnaldia glomerata, *Agardh.*

Hab. In slow streams and ditches, adhering to stones, sticks, &c. Not common in Scotland; more common in England and Ireland. I have got it in several places in Ayrshire.

Stem round, branched; ramuli in tufts, which are frequently alternate, and always ciliated; tufts divergent.

The first time I ever saw it was at Lochranza, in the island of Arran, in a little limpid pool in a stream from the mountain. I found great difficulty in catching it; it was so lubricous that it slipped through my hands like an eel, and so fragile that when caught it broke by its own weight. When first removed from the water, it is like a mass of coloured jelly without form or organization; but when placed again in water to be spread on paper, it unfolds itself very beautifully.

The *Draparnaldiæ* are universal favourites, the colour

and the structure being so lovely. *D. glomerata* is the largest of the family, and when found in a young state before the zoospores have escaped, it is truly beautiful. It adheres closely to paper, but it is so gelatinous than in most cases it must be allowed to dry on the paper before it is pressed, as it adheres to whatever covers it.

2. DRAPARNALDIA NANA, *Hassall*. (Plate XX. fig. 79, plant, natural size; on the right, a branch magnified, and underneath, a branchlet still more magnified.)

Filaments highly mucous, very slender, sparingly branched. Branches acuminate, not usually ciliated. Cells rather broader than long.—*Hassall*.

Hab. In streams, adhering to grass and weeds, and sticks and stones. Mr. Hassall says that is not uncommon in England in spring. It is rare in Scotland.

The first time I found it, it was adhering to a piece of wood in a runlet of water pumped from a coal-pit near Stevenston, where the obstructing wood and stones formed a little waterfall. The next time I got it in great beauty in October, attached to withered grass, which, though rooted on the bank, was in part floating in the stream of Stevenston burn, at a place generally affected by the tide. For seven years I never saw it again, though I often sought it at the same place at the same season. In May 1848, however, I got it in great abundance and beauty in the same rivulet attached to pond-weed. The tufts, waving gracefully in the stream, were two or three inches in length, and, when cautiously handled, could be brought out entire. Aided by my youngest daughter (who did not much like the wading, as the little flounders were always pouncing upon her feet) I procured a great number of specimens.

D. nana is very like *D. plumosa*, but Mr. Hassall thinks it quite different.

3. DRAPARNALDIA ELONGATA, *Hassall.*

This, which is regarded as a new species, was first found by Major Martin in a quarry pool near Saltcoats. It was got by D. L. in April 1849, in a slow-running stream at the Turf Dyke coal-pit, Stevenston. It was lying at the bottom, growing on weeds and stones, in inconsiderable masses. The filaments were slender, and could not support much of the mass when an attempt was made to lift it out of the water. It bears a considerable likeness to *D. nana*, though the green is lighter, and the filaments longer. Mr. Hassall says that it is rare; that he has only twice got it,—once growing in a horse-trough near Cheshunt.

ZYGNEMA, *Agardh*.

Gen. Char. Filaments articulated, simple, finally united in pairs by numerous transverse tubes. Endochrome consisting of granules arranged in spiral rings, or in a simple row, which, after conjugation, are condensed into a globule in one of the filaments, or in the transverse tubes.—The name is from the Greek words signifying a *yoke* and a *thread*, the threads, though at first separate, being afterwards yoked together.—*Harvey*.

Zygnema is one of the genera of the order *Conjugateæ*. The *Conjugateæ* are undoubtedly the most curious tribe of *Confervæ*. Their filaments are simple, and of uniform diameter. They are mostly unattached, and, being the inhabitants of stagnant waters, are in no danger of being disturbed in their curious process of fructification. The simple filaments are composed of elongated cells, placed end to end, and held together by an enveloping membrane. The interior of these cells is occupied chiefly with endochrome, sometimes like stars, spirally arranged, and at other times filling the cavity of the cells. What follows is very remarkable.

When the filaments are fully grown, as they are in close juxtaposition to each other, the cells are observed to send forth little conical processes or tubes, which unite with similar protrusions from corresponding cells of an adjoining filament, thus establishing a passage of communication betwixt the cells. In the meantime, if the plant be a *Zygnema*, the endochrome in the spiral tubes becomes confused, and the contents of one cell pass through the connecting tube, and mingle with those of the other, forming a circular or oval body, of a dark green colour. It is remarkable that the cells of one part of the filament will part with their contents and remain empty, while in another part of the same filament, they will receive the contents of the cells of another adjoining filament.

Some of the species of the genus *Zygnema* do not thus unite with other filaments; the round dark-green granular balls being in these formed by the union of the contents of two adjoining cells in the same filament. When two cells are thus conjugating, the cell which has the greater portion of the matter receives the contents of the other. In the course of a few days, the *sporangia*, or globules, are formed, and are invested with two or three membranes, to preserve, we doubt not, the vitality of the seed. In a week or two the filaments separate, and break down at the joining of the cells, and the zoospores bursting from them are disengaged and fall to the bottom, to spring up after the rigour of winter, or even during winter, when there is not a continuance of frost.*

* See a fuller and better statement of these matters in Hassall's excellent work on British Fresh-water Algæ.

1. ZYGNEMA QUININUM, *Agardh.*

The filaments are pale yellowish-green; the spires perform three revolutions in each cell; the spores are simple. Very common in ditches and pools, in cloudy masses, of a pale green colour; filaments glossy, and marked with a spiral line resembling a constant repetition of the Roman numeral V., or five, whence the specific name *quininum*. This is the chief distinguishing mark from another very common species, which has two spiral lines crossing each other, and thus repeating throughout its whole length the Roman numeral X., whence it is called *Zygnema deciminum*. This is well represented in Pl. XXIII. figs. 3 and 4, of Mr. Hassall's work, while *Zygnema quininum* is figured in Pl. XXVIII. figs. 1 and 2.

DESMIDIEÆ.

The character of this family is thus given by Mr. Ralfs:—"Fresh-water, figured, mucous, and microscopic Algæ, of a green colour; transverse division mostly complete, but in some genera incomplete; cells or joints of two symmetrical valves, the junction always marked by the division of the endochrome, often also by a constriction; sporangia formed by the coupling of the cells, and union of their contents." This description is taken from an admirable book by Mr. Ralfs, of Penzance, on British Desmidieæ,—a work of great research, and illustrated by many beautiful plates. I shall not attempt to lead my young friends into the depths of microscopic Algology, yet, in touching at all on freshwater Algæ, I could not refrain from giving a slight notice of this exceedingly interesting tribe. They are all inhabitants of fresh water. Their colour is green, with the exception of a few of one genus, whose outer integument is coloured, though the internal matter is green, while their most obvious peculiarities are the beautiful variety of their forms, and their external markings; their most distinguishing characteristic, as Mr. Ralf observes, is the evident division into two valves, or segments. Each cell, or joint, of the *Desmidieæ* consists of two similar valves or segments, and the line of junction is in general well marked. Mr. Ralfs is decidedly of opinion that the two valves are but one cell, differing on this point from any other writers on the subject, except Professor Kützing, who in his 'Phycologia Germanica' has arrived at the same conclusion, by independent observations. Mr. Ralfs states that the multiplication of the cells by repeated transverse divisions is full of interest, both as it relates to themselves, and in the remarkable manner in which it takes place, and because it unfolds the nature of a process in other families; and furnishes a valuable addition to the knowledge of their structure and physiology. The process is very evident in the genus *Euastrum*, for, though the frond is really a single cell, yet in all its stages it appears like two, the segments being always distinct, even from the commencement, being separated from each other by the length of the connecting tube, which is converted into two hyaline lobes. These lobes, increasing in size, acquire the colour, and gradually put on the appearance, of the old portion. Of course, as they increase, the original segments are pushed further asunder, and at last are disconnected, each new lobe taking with it an old segment to supply the place of that from which it was separated, so that every new specimen of *Euastrum* is partly new and partly old. A single glance, however, at Mr. Ralfs's Plate XI. fig. 2, *Euastrum verrucosum*, will give a better idea of this than all the words we can employ.

MICRASTERIAS, *Agardh*.

Gen. Char. Fronds simple, lenticular, deeply divided into two lobed segments; the lobes inciso-dentate (rarely only bidentate) and generally radiant.—*Ralfs*.

1. MICRASTERIAS DENTICULATA, *Brébisson*. (Plate XX. fig. 80, a mature frond, and under it a dividing frond.)

Hab. Penzance, Mr. Ralfs; Kent, Mr. Jenner; Henfield, Mr. Borrer; Bristol, Mr. Broome; Ambleside, Mr. Sidebotham; Aberdeenshire, Dr. Dickie; Stevenston, in Ayrshire, D. L.

This species is not uncommon. It is found in marshy ground and in ditches. The frond is large and nearly circular, each segment is five-lobed; the colour is bright green; frequently the margin of the frond is colourless. Originally the two segments of the frond are united by a narrow tube, as seen in the plate. This connecting tube lengthens, expands, and becomes two young segments, of a lighter green colour, as in the lower figure. When these two segments have become full-sized, they separate, and form two fronds, of each of which one half is old, and the other half new.

HYALOTHECA, *Ehrenberg*.

Gen. Char. Filament elongated, cylindrical, very gelatinous; joints having either a slight constriction, which produces a crenate appearance, or a grooved rim at one end, which forms a bifid projection on each side; end view circular.—*Ralfs*.

1. HYALOTHECA DISSILIENS, *Smith*.

Hab. Penzance, &c., Mr. Ralfs; North Wales, Mr. Borrer; Kent, &c., Mr. Jenner; Essex, Mr. Hassall; Bandon, Professor Allman; Ambleside, Mr. Sidebotham; Bristol, Mr. Broome; Stoke Hill, Mr. Thwaites; Aberdeen, Dr. Dickie; Ayrshire, D. L.

I have fixed on this species because I do not suppose that it is uncommon. I have got it in several localities; in Ayrshire, in particular, it is very abundant, in a little pond near Turf Dyke coal-pit, Stevenston. It is very beautiful when examined with a microscope, or even a good pocket lens. I first knew it under the name of *Conferva dissiliens*, and in its general appearance it greatly resembles some of the *Confervæ* proper, unbranched. The filaments are of considerable length, and fine as a human hair. Look at one of the filaments: though you see no difference in the two extremities, you would be disposed to say that one end must be the base and the other end the summit—the former the older, and the latter the younger of the two. You are fairly out; for the two ends are the oldest portions of the filament. At first there were two valves, forming one cell, constituting the plant. These two segments or valves were connected by a tube, which lengthened as we have seen in *Micrasterias*, and gradually formed two new valves. These grew and parted asunder the two original valves, and the new, joining with the old, formed two cells instead of one. This process is repeated till a filament is formed, several inches in length, and consisting of numerous cells; the two segments at the end of the filament, though now far separated, being the two valves or segments of which the original cell consisted.

Mr. Ralfs has a long and able dissertation on a disputed point, whether the *Desmidieæ* are animal or vegetable, and proves very satisfactorily that they are vegetable.

In conclusion, we may advert to a curious fact, mentioned by Mr. Ralfs, of *Hyalotheca*, and other *Desmidieæ*, being found growing in an old water-butt, in water derived from the clouds alone. Did these plants descend from the clouds? Yes, but they had previously ascended from the earth. The "High and Lofty One" disdains not to provide for the continuance and wide diffusion of what He has in wisdom made. To the seeds of many land-plants He gives what serves all the purposes of wings. To the spores of many minute aquatic plants He gives, by means of *cilia*, the power of waving to and fro in the waters, till they find a suitable place for rest and growth. They are so light that they may be raised into the atmosphere during the process of evaporation, and driven about by the slightest breath of wind.

Directions for Collecting, Spreading out, Preparing, and Preserving Sea-weeds.

Much of the pleasure and much of the benefit arising from the study of Algology consists in the pursuit. It is in so far like hunting and fishing; there is all the excitement of hope, and all the advantage of exercise; and there is this in its favour, that, however great the success, there is no life taken, no blood shed, and the subsequent enjoyment is not limited to a short repast, but may be continued for many years.

Let the young Algologist provide a tin vasculum, or an oil-skin bag, in which he may deposit his marine stores. As some of the finer Algæ soon fade in colour when exposed to the air, it may be well to have a small wide-mouthed flask in which they may be carried home floating in sea-water. A staff with a crooked head is not a bad accompaniment. Thus accoutred, let him proceed to the shore at ebb-tide, and examine the *rejectamenta* cast out by the sea, turning them over with his staff, that no newly-buried beauty may be allowed to perish. Let him then carefully examine the Algæ growing on the uncovered rocks, extending the investigation to those rocks or stones that are still partly under the waves. When the tide is turned, and begins to flow, fine weeds may often be got floating in little bays, or where currents betwixt rocks are formed. These may be very easily caught by the weed-gatherer's staff.

When the vasculum is filled, or the time is up, or the collector tired, let the spoils of the sea be carefully examined when he reaches home. There will be much uncertainty in many cases as to the contents of the vasculum, or oil-skin bag, till they are floated and spread out on paper. Then it is that there is scope for fine taste, and for the delicate manipulation of ladies' fingers; nature must be consulted as the sure instructress for laying out the specimens in the most graceful manner. Place on the table a basin of fresh water, to cleanse the weeds from sand or any impurity. Let only a small portion of the mass at a time be put into the basin, as many species begin to decompose when placed in fresh water. If the specimens are of large size, they may, after cleansing, be floated in a shallow tin-tray filled with fresh water; but if they are only of moderate size, a white soup-plate will answer the purpose: let the plate be nearly filled with luke-warm water. Let a good supply of paper be at hand; and, as much of the beauty of the specimen depends on the quality of the paper, it should be fine, and at the same time stout, almost as good as drawing-paper. The paper should be cut so as to be quite smooth at the edges, and as this is best effected by the bookbinder's knife, it is well to have it done when the paper is purchased.

Having got the paper neatly cut into square and oblong portions, of different sizes, take a piece suited to the size of the specimen, and place it under the weed floating in the water, then putting the left hand under the paper, bring it near the surface, and gently move the sea-weed till it assume on the paper a natural and graceful form; the fingers of the right hand may be employed in helping to arrange the branches of the plant, or some sharp-pointed instrument may be used for this purpose,—a penknife, the quill of a porcupine, or, what is still better, as being less sharp, the point of a silver fruit-knife. A pair of nice little scissors should be at hand, to remove any superabundant branches.

When the specimen is properly arranged, let the paper on which it is spread be very cautiously removed from the water, for if the position of the plant is changed the work may require to be done over again. When the specimen is removed from the water, it may be placed for a little while in a sloping position, to allow the water to run off, and during this time other specimens may be treated in the same way.

The drying and pressing processes then begin. Before any part of the paper is completely dry, place the specimens on several folds of blotting-paper, quarto size; and cover them with a fold of muslin, and over the muslin lay several folds of blotting-paper, repeating this operation till all the specimens that have been laid down are covered with a fold of muslin, and several folds of blotting-paper. If a screw-press is at hand, let the whole be placed in it and *gently* pressed. Strong pressure at first would bruise the plant, especially if at all gelatinous. After some hours of slight pressure the whole may be removed, and either treated with a fresh supply of muslin and paper, or those in which they were may be dried before the specimens are again placed in them. The advantage of being covered with a fold of muslin will then appear; for in general none of them will be found adhering to the muslin, whereas had they been covered only with blotting-paper the half of the specimens would have been spoiled by adhering to it.

The whole may then be replaced in the press, and considerably stronger pressure applied to them, and under this they may be allowed to remain for a day and a night. In shifting them the second time the muslin coverings may be removed. When permitted to remain till the plants are quite dry there is danger of their leaving chequered impressions on the specimens. They may then be replaced in the press, and very strong pressure applied. They should be shifted once a day for a week, and the paper dried; and at the end of that time they may be deposited in the herbarium, when they will be found adhering so closely to the paper as to have all the appearance of a beautiful painting.* Where there is not a press, the want is easily supplied. All that is necessary is two boards the size of the blotting-paper, and three weights of stone or cast-iron. The blotting-paper containing the specimens being placed betwixt the boards, one weight may be placed above them at first, two at the

* A learned Professor going abroad asked me to give him some specimens connected with natural history for the museum of his college. I sent him an *album* filled with sea-weeds, and he wrote to me that he was exceedingly obliged to me for the beautiful *sketches* done by my daughters!

second shifting, and all the three afterwards, and let the last be a very heavy one. When the specimens are taken out of the blotting-paper, before they are placed in the herbarium, the scientific name, the locality and the date, should be neatly written at the bottom.

Though what we have said respecting laying down and pressing is suitable for plants in general, there are exceptions; there are some that may be treated in the way we have prescribed, and yet they would not adhere. If they are cartilaginous, or coriaceous, or destitute of gelatine, however firmly pressed, they will not cling to paper; these require a little gumming, or a little isinglass, which leaves not a glare like gum.

Some delicate plants, that lose their fine colour when prepared in fresh water, retain it considerably better when they are floated in sea water, or, if this is not at hand, in water in which there is a solution of common salt.

The Great Tangle, and some of the large *Fuci*, which are not pliant, and which lose their colour, becoming black in the herbarium, are by Algologists that we know dipped in hot water for a little, which both renders them more pliant, and prevents them, for a time at least, from becoming black. The natural colour might perhaps be in some degree preserved if, according to the practice of other Algologists, a coat of varnish were given them before they are placed in the herbarium.

It is a general complaint that the *Porphyræ* do not adhere to paper in drying, but shrink and become torn, in consequence of starting from the paper. This is very easily prevented. Let them be spread out and covered with muslin in the common way, but let not the muslin be removed for two or three days; yet, though the muslin is to be allowed to remain, let them have, along with the other species, a supply from time to time of blotting-paper, not only dry, but heated at the fire. In four or five days they will be quite dry, and they will adhere so firmly to the paper that they will seem a part of it.

Very gelatinous kinds, such as *Gloiosiphonia, Mesogloia, Batrachospermum, Draparnaldia*, &c., would in some cases be destroyed, if covered and pressed in the manner we have directed. They would be bruised by the pressure, and would adhere to the muslin or paper, however lightly laid over them. They must be allowed to lie exposed to the air till they are dry, and then, after moistening with a sponge the under side of the paper, strong pressure may be applied to them. But, though these cautions are in general necessary, there are certain states of these plants in which they may be covered and pressed in the common way, and those that are thus treated make by far the finest specimens.

There are several ways in which collections of sea-weeds may be preserved after they have been carefully prepared. They may be kept loose betwixt folds of paper, and the sheets may be arranged alphabetically according to the names of the plants they contain. This is found very convenient when reference is required, or when a selection from them is to be made.

When the student of Algology has got specimens named on good authority, they should be kept separate, and may be attached by fine pins to the sheets in which they are placed. This in their case is preferable to any permanent fastening, because it may often be necessary to examine their structure and fructification by placing them under the microscope, or by holding them up betwixt the eye and the light that they may be examined with the aid of a good lens.

When specimens are placed in an album, slits may be made to receive the four corners, and in this way they may easily be removed and replaced at will. If the collection is chiefly valued for its beauty—and few things are more beautiful than a good collection of well-prepared Algæ—a handsome album should be procured, formed of stout coloured paper, and on the pages of this album the specimens should be tastefully arranged according to their size and form, and then they may be made to adhere to the strong coloured paper by touching the under side of each corner with well-made paste; or, if there is no wish to remove them, by applying the paste with a camel-hair brush to the whole of the under side. After a short pressure, they will adhere in the firmest manner, and, judging from what I have lately seen, there is no way in which a fine collection appears to so great advantage.

Having thus given the best directions I can for causing specimens firmly to adhere to paper, that they may be placed in the herbarium, it would not be right to conclude without mentioning that there are other ways of preparing Sea-weeds for ornamental work, and also for the herbarium. Mrs. Hunter, of Drum, near Edinburgh, is most successful in preparing them for ornamental fancy-work, which she disposes of for charitable purposes. She kindly gave me a most beautiful specimen of her work, along with a detail of her mode of procedure; but as I did not succeed in reducing her instructions to practice, I fear that I do not sufficiently understand them to be a good instructor of others.

I have lately corresponded with Mr. M. C. Pike, Pool Valley, Brighton, who prepares for sale fancy-work composed of Sea-weeds, and also specimens for the herbarium. They are free—or, what he calls, transferred—from the paper, and when they are laid on a clear page, they are not gummed to it in any way, but merely retained in the position in which they are placed by very small cross slips of paper, as is often done with dried specimens of flowering plants. I learn that he has so great a sale that he can with difficulty supply the demand, and I do not wonder at it, for I could not have imagined that Sea-weeds could, in this unattached way, have been so beautifully and tastefully laid out. His mode of preparing them is a secret, and we doubt not it will yield him a good remuneration for his great artistic skill.

We doubt not that those who engage in the study of Algology will thank us for the following instructions as to the mode of preparing the fructification, &c., as objects for the microscope.

From the minuteness and delicacy of their structure, it will at once be perceived that the aid of a good microscope, or at least a powerful lens, is an indispensable requisite, in prosecuting the study of the Algæ in its scientific details. The forceps, knives, and scissors, used in dissecting other vegetable tissues, will do equally well for them; and, as most of them are not of so perishable a nature as forms of a higher rank, any delay, from other engagements, in mounting them, is not of material consequence.

When mounted on glass slides, or other modes in which they can be viewed by transmitted light, they form permanent objects, not only of scientific interest, but also, from the beauty of their form, of agreeable entertainment to even the uninitiated, and it is to several simple plans of preparing them for this purpose that we devote the few following paragraphs.

In the first place, we must mention the few implements and materials required in preparing the cells, and afterwards mounting the objects in them. These are, slips of crown or patent plate glass, rough or smoothed at the edge. The size chiefly used is that approved by the Microscopical Society, viz., three inches by one inch. In using a uniform size, great facility is given in the way of exchange.

Thin or microscopic glass, cut into circular or square pieces of various sizes, as covers for the objects immersed in fluid.

Phials with a supply of either of the liquid preparations enumerated below.

Japan varnish, gold size, and thin pieces of gutta percha, or other materials used for forming the cell in which the object is to be preserved in a fluid state.

Knives, forceps, and scissors, of various constructions.

Saucers with and without lips, watch-glasses, hair-pencils, chamois leather for cleaning glasses.

Having these ready, with the help of a hair pencil a square or circular space is marked off with varnish, exactly on the centre of a glass slide, forming a narrow band or cell, an eighth of an inch or so in diameter. The varnish in this process should be laid on very thin, and a second coat should not be given till the first is quite dry. A third or fourth may be added according to the depth required for the specimen. In making the gutta percha cell, we will suppose a portion of that material cut out with a knife or punch, of the same size and form as that mentioned for the one of varnish. With this ready, place a glass slide on the hob, or other stand at the fire-place, till it is of such heat as will barely allow of handling. A pair of forceps will hold it on one side, while with the other the gutta percha is dropped on the centre, and pressed gently down with some flat substance. When quite cooled, the cell may be further cemented to the glass by several layers of varnish applied to the exterior edge.

We will now suppose the collector to have returned from the shore, laden with the spoils of a recent storm, in the shape of some finely-fruiting specimens of the various species of *Delesseria, Polysiphonia, Callithamnion,* or allied genera. After supplying himself with a sufficiency for drying, he will have as many fragmentary portions left as will make a number of preparations.

He is now ready to try his hand at mounting these, and we proceed to give briefly a few directions, referring such as would be proficient in the art for further details to Quekett's admirable work on the microscope, and also articles on the subject in the Annals and Magazine of Natural History for February and April 1845. The first mode we mention is one to which we have not yet referred, inasmuch as no barrier cell requires to be raised for confining the fluid. Some of the Callithamnions, and allied genera, whose delicate branchlets are thin as the "web of the gossamer loom," require no cell with raised walls. All that is necessary is to touch one of the thin glass covers—previously carefully cleaned—all round with varnish. The delicate object having been laid on the surface of the glass slide, with the necessary quantity of preserving fluid, the glass cover is carefully dropped over it. As the varnish repels the water, none but superfluous liquid escapes from beneath, which can be removed with small pieces of blotting-paper. After standing for a little, the varnish gets firmer, and the object may be sealed up as mentioned in the next mode.

In using the varnish cell, having ascertained that the walls are quite dry, their surface even, and the enclosed space free from dust, we may proceed to prepare an object. A thin glass cover having been selected, a slight degree smaller than the exterior edge of the cell to be used for cementing the cover, the preserving fluid, with an object suitably proportioned in size and thickness, is laid out on the centre, and the glass cover thereafter carefully laid on. It is better to have rather too much than too little liquid, as what remains can be sucked up with blotting paper, and the operator must not be disappointed if he fails frequently in excluding the air. In a short time it is ready to seal up, by giving over the edges of the thin glass a thin coating of varnish. After standing on a dead level for twenty-four hours, a second and third coat should be given; and if the colour of the Japan varnish is not liked, a coat of sealing-wax varnish will make a nice finish.

The gutta percha cells are used much in the same way; the surface of the cell getting a slight coating of varnish before putting in the specimen, though if the cell be quite flat this part may be dispensed with. Space will not permit to speak of the paper covers, modes of attaching names to objects, and packing the slides in cases or boxes, which all require attention. All that we would say about preserving them is to see that they are thoroughly dry, and kept without touching each other, in order to secure their safety.

Larger specimens of the fruited fronds of the *Fuci*, and other *Melanospermeæ*, should be preserved in phials.

Liquid Preparations for preserving Algæ.

"Mr. Thwaites's Creosote preparation."

1 part alcohol; 14 parts water. To be accurately filtered through creosote.

This should be filtered through prepared chalk, and the solution allowed to stand some time before use.

"Goodby's Solution for Marine Algæ."

4 oz. bay-salt; 2 oz. alum; 4 grains corrosive sublimate; 2 quarts boiling water.

Some use simple spring water, or sea water, which answers well enough for some.*

* For the preceding instructions for preparing microscopic objects, we are indebted to Mr. R. M. Stark, 2, Hope Street, Edinburgh, who keeps a good assortment of them for sale.

GLOSSARY.

THE FROND.

Frond, is employed to signify all parts of a sea-weed except the root, and, in some cases, the stem, when it is very distinct from the other parts of the plant.

The Frond may be

Bifid, cleft into two segments; *Bilobed*, divided into two lobes; *Dichotomous*, regularly and repeatedly cleft in two, as in *Dictyota dichotoma*; *Furcate*, forked.

Capillary, slender, hair-like, as in the *Confervæ*.

Compressed, flattened laterally, as in *Ent. compressa*.

Constricted, drawn together as if tied, at intervals, as in *Chorda lomentaria*.

Continuous, without interruption, prolonged.

Convolute, rolled together; *Involute*, rolled inwards; *Revolute*, rolled backwards.

Cordate, heart-shaped at the base; *Obcordate,* heart-shaped at the apex.
Crenated, notched.
Cuneate, wedge-shaped.
Cylindrical, round and elongated.
Denticulated, toothed.
Ensiform, sword-shaped.
Falcate, sickle-shaped.
Filamentous, slender, thread-like.
Filiform, string-like, the size of common twine.
Fimbriated, fringed.
Flabelliform, fan-shaped.
Flexuous, bending to one side or the other, wavy.
Geniculated, bent like the knee.
Hastate, shaped like a spear.
Laciniated, cleft more or less deeply.
Lanceolate, shaped like a lance.
Lenticular, circumference round, surface depressed above and below.
Linear, narrow, the same width all along.
Linguliform, tongue-shaped.
Ovate, rounded at the base, tapering towards the apex, egg-shaped.

Obovate, the preceding reversed.
Oval, or *Elliptical,* equally rounded at both ends, the length exceeding the breadth.
Palmate, shaped like the hand with the fingers extended.
Pinnatifid, cut transversely into oblong segments.
Plane, level, flat.
Proliferous, when a second frond springs from the first.
Reniform, kidney-shaped.
Saccate, in the form of a bag.
Simple, undivided, unbranched.
Spathulate, rounded at the apex, and tapering at the base.
Tubular, hollow, round like a tube.

The Frond in *Substance* may be

Carnose, of fleshy consistence.
Cartilaginous, stiff, gristly.
Coriaceous, leathery, tough and elastic.
Corneous, horny.
Flaccid, soft, collapsing when removed from the water.
Gelatinous, jelly-like.
Rigid, harsh and stiff.

In *Structure,* the Frond may be

Cellular, when composed of small cells.
Filamentous, when made up of threads.

2 c

Gelatinoso-cartilagionus, betwixt gelatinous and cartilaginous.

Gelatinoso-membranaceous, betwixt gelatinous and membranaceous.

Punctated, dotted.

Reticulated, veined like net-work.

Striated, streaked with lines.

The smaller divisions of the Frond, called *Ramuli,* branchlets, or *Ramelli,* little branchlets, may be

Appressed, approaching the stem or branch, so as to be almost in the same direction.

Articulate, jointed.

Byssoid, forming tufts of slender filaments.

Corymbose, level-topped; the branchlets of different length, but level or nearly so at the top.

Distichous, placed in two opposite rows.

Divaricated, when the direction is between the patent, or spreading, and horizontal.

Fasciculated, tufted and level-topped.

Imbricated, overlapping each other like tiles.

Inarticulate, not jointed.

Multifid, much divided, or cleft.

Patent, spreading.

Pectinate, with the divisions like the teeth of a comb.
Pectinato-pinnate, partaking of both the preceding characters.
Pinnate, winged.
Quadrifarious, arising from all sides of the branch.
Setaceous, bristly.
Secund, when the branchlets bear another series on one side.
Verticillate, whorled, set in a circle round the stem.

The Fructification.

Capsules, small pitcher-shaped bodies containing spores, sporules, or seeds.

Cilia, eye-lash like bodies, with which the spores of many of the Algæ are clothed.

Endochrome, a dark-coloured mass in the filaments that forms the seed.

Granules, seeds in the form of little grains, collected in patches on the frond, called sori, or placed in the tips of branchlets. Capsular seed is called *primary,* and granular seed *secondary,* only by way of distinction, for they are equally productive of plants.

Involucre, a small cover of fruit, formed, as in *Ceramium rubrum,* of short branchlets.

Pericarp, the cover of the seed or fruit.

Receptacles, variously shaped bodies containing the seeds.

Sori, as we have already said, are patches either defined or scattered, in which the granular seed is placed.

Sporidia, clusters of sporules.

Stichidia, Antheridia, Coccidia, Ceramidia, Utricles, Favellæ, different kinds of fructification, explained in Chapter V. page 30.

Tetraspores, a mass of four spores conjoined.

Tubercles, small round masses generally containing seed.

Colour.

When there is scarcely any colour, and the parts are almost transparent like glass, the plants are said to be *hyaline, diaphanous, pellucid.* The opposite of this is opake.

Cinereous, ash-coloured.

Ferruginous, rust-coloured.

Fuliginous, smoke-coloured.

Fuscous, reddish brown.

Glaucous, mixture of green and blue.

Iridescent, having the colours of the rainbow.

Olivaceous, a dusky green, inclining to brown.

This is at times very remarkable in *Chondrus crispus,*

which we saw lately having a rich metallic lustre, the tints being as brilliant as those in the neck of the peacock. We have also seen *Nitophyllum laceratum* very iridescent.

Other terms not comprised in the foregoing.

Abortive, not reaching perfection.
Abnormal, contrary to the regular order of growth.
Accessory, ramuli, differing from the ordinary branchlets, and for a special purpose.
Acuminated, with a long tapering point.
Aculeated, pointed like a prickle.
Adnate, adhering to an object by the whole surface.
Agglutinated, glued together.
Aggregate, collected, or grouped together.
Amorphous, without regular shape.
Anastomose, to grow into another body and unite with it.
Areolated, marked like a pavement.
Annulated, ringed.
Axillary, in the angle called the axil, formed by the junction of branch and stem.
Axis, central portion of the frond.
Basal, at the base.
Bullated, blistered.
Caulescent, having a stem.
Cirrhose, with tendrils.

Clavate, shaped liked a club.
Conceptacle, a hollow case, containing spores.
Costate, ribbed.
Cryptogamous, plants not having flowers.
Deciduous, falling off.
Deflexed, bent downwards.
Dichotomous, branched by repeated forkings.
Disc, surface of a frond within the margin; also, the flat base by which many Algæ adhere to rocks.
Dissepiments, the partitions of the articulate Algæ.
Falcate, sickle-shaped.
Fastigiate, when the branches are parallel and pointing upwards.
Flexuous, bent from side to side.
Frondlet, a little frond.
Frustules, the joints of which dichotomous plants are composed.
Fusiform, spindle-shaped.
Gibbous, the surface elevated at a particular place.
Glandular, having glands, bodies containing juices.
Habitat, the place of growth where a plant is found.
Involute, rolled inwards.
Lateral, at the sides.
Limbus, a border.

Mammillated, hemispherical, with a wart on its tip.
Midrib, a large vein, termed percurrent when continuing the whole length of the frond.
Mitriform, mitre-shaped.
Moniliform, beaded like a necklace.
Mucronated, rounded apex armed with a spine.
Nerve, a faint vein.
Obsolete, when wearing away.
Parasite, Parasitic, growing on another plant.
Patent, spreading.
Pedicel, Peduncle, the stalk of the fruit.
Periphery, an envelope.
Phenogamous, flowering plants.
Pinnæ, winged leaflets or portions of the frond.
Placenta, a pillar often formed by the thickening of the partitions of the pericarp: the part to which the spores are attached.
Polymorphous, many-shaped.
Process, any projecting part.
Pulvinate, shaped like a cushion.
Pyriform, pear-shaped.
Retiform, like net-work.
Revolute, rolled back.
Rotund, round.

Scutate, shaped like a shield.
Segments, divisions of the frond.
Septa, bands, partitions.
Serrated, like a saw.
Sessile, having no stalk.
Sinuous, when the margin has numerous, shallow, blunt indentations.
Stipitate, having a stem or stalk.
Sub-rotund, roundish; the prefix meaning that the character does not strictly apply.
Subulate, awl-shaped.
Terete, round, cylindrical.
Tortile, twisted.   *Voluble*, twining.
Truncated, cut across.
Umbilicated, the surface depressed, surrounded by an elevated margin.
Uncinated, hooked.
Urceolate, expanded at both ends, and contracted in the middle: shaped like an ancient pitcher.
Vesicle, a bladder.   *Utricle*, a little bladder.
Virgate, long and straight, like a wand.
Whorled, surrounding a branch in a ring.
Zigzag, angularly bent from side to side.

ALPHABETICAL LIST

OF

SPECIES.

Marine Algæ.

	Page.		Page.
ALARIA esculenta		Callithamnion arbuscula	176
(Plate I. fig. 4)	116	—— barbatum	179
Arthrocladia villosa	134	—— brachiatum	180
Asperococcus compressus		—— Brodiæi	177
(Plate V. fig. 18)	147	—— byssoideum	180
—— echinatus	149	—— corymbosum	
—— Turneri (Pl. II. fig. 7)	148	(Plate VI. fig. 23)	183
Bangia fusco-purpurea	325	—— cruciatum	175
—— ? ceramicola	326	—— Daviesii	187
Bonnemaisonia asparagoides		—— floccosum (Pl. VI. f. 22)	178
(Plate XII. fig. 45)	266	—— floridulum	185
Bostrychia scorpioides		—— gracillimum	
(Plate XI. fig. 41)	288	(Plate VI. fig. 21)	182
Bryopsis hypnoides	310	—— Hookeri	177
—— plumosa (Pl. XVI. f. 64)	309	—— pedicellatum	185

	Page.
Callithamnion pluma	179
—— plumula	175
—— polyspermum	181
—— roseum (Pl. VII. f. 26)	178
—— Rothii	185
—— sparsum	186
—— spongiosum	184
—— tetragonum	180
—— tetricum (Plate VII. fig. 25)	177
—— thuyoideum	183
—— tripinnatum	181
—— Turneri	179
Calothrix confervicola	332
Carpomitra Cabreræ	133
Catenella Opuntia (Plate X. fig. 39)	217
Ceramium acanthonotum (Plate X. fig. 38)	198
—— botryocarpum	202
—— ciliatum	198
—— decurrens	200
—— Deslongchampsii	201
—— diaphanum (Plate VII. fig. 27)	197
—— echionotum	199

Ceramium fastigiatum
—— gracillimum
—— nodosum
—— rubrum
    (Plate VIII. fig. 29)
Chondrus crispus
    (Plate IX. fig. 33)
—— Norvegicus
Chorda filum (Plate III. fig. 9)
—— lomentaria
Chordaria divaricata
—— flagelliformis
Chrysimenia clavellosa
    (Plate XIV. fig. 56)
Chylocladia articulata
    (Plate XIV. fig. 53)
—— kaliformis
    (Plate XV. fig. 57)
—— ovalis (Pl. VIII. fig. 31)
—— parvula
—— reflexa
Cladophora arcta
    (Plate XVII. fig. 66)
—— gracilis
—— Hutchinsiæ
—— lætevirens

INDEX.

	Page.
adophora lanosa	
(Plate XVII. fig. 65)	314
pellucida	314
rectangularis	
(Plate XVII. fig. 68)	313
refracta	317
rupestris	
(Plate XVII. fig. 67)	313
adostephus spongiosus	154
verticillatus	
(Plate IV. fig. 14)	154
dium adhærens	307
amphibium	307
Bursa	307
tomentosum	
(Plate XVI. fig. 63)	307
erva ærea	319
Linum	319
Melagonium	
(Plate XVIII. fig. 70)	318
tortuosa	
(Plate XVIII. fig. 69)	318
rallina elongata	277
officinalis	
(Plate XIV. fig. 54)	276
squamata	277

	Page.
Crouania attenuata	207
Cruoria pellita	215
Cutleria multifida (Pl. II. f. 6)	136
Cystoseira ericoides	103
Dasya arbuscula	305
—— coccinea	304
—— ocellata	305
—— venusta	305
Delesseria alata	257
—— angustissima	258
—— Hypoglossum	259
—— ruscifolia (Pl. XIII. f. 49)	260
—— sanguinea	
(Plate XIII. fig. 50)	252
—— sinuosa	255
Desmarestia aculeata	131
—— ligulata (Plate I. fig. 3)	129
—— viridis	130
Diatoma marinum	335
Dictyosiphon fœniculaceus	144
Dictyota dichotoma	
(Plate II. fig. 5)	141
Dudresnaia coccinea	207
—— divaricata	208
Dumontia filiformis	
(Plate X. fig. 40)	229

	Page.
Ectocarpus brachiatus	163
—— crinitus	161
—— distortus	161
—— fasciculatus	160
—— granulosus	162
—— Hincksiæ	160
—— Landsburgii	161
—— littoralis	159
—— longifructus	160
—— Mertensii	163
—— pusillus	161
—— siliculosus (Plate V. fig. 19)	159
—— sphærophorus	162
—— tomentosus	160
Elachistea curta	167
—— flaccida	167
—— fucicola	167
—— pulvinata	167
—— scutulata	167
—— stellulata	167
—— velutina	167
Enteromorpha compressa (Plate XVI. fig. 62)	328
—— erecta (Plate XVIII. fig. 71)	328
Enteromorpha intestinalis (Plate XVIII. fig. 72)	
Exilaria fulgens	
Fucus canaliculatus	
—— ceranioides	
—— Mackaii	
—— nodosus	
—— serratus (Plate I. fig. 1)	
—— vesiculosus	
Furcellaria fastigiata	
Gelidium cartilagineum	
—— corneum	
Gigartina acicularis (Plate XI. fig. 42)	
—— mamillosa	
—— pistillata	
—— Teedii	
Ginnania furcellata	
Gloiosiphonia capillaris	
Gracilaria compressa	
—— confervoides (Plate XI. fig. 44)	
—— erecta	
—— multipartita	
Grateloupia filicina	
Griffithsia barbata	

	Page.		Pa
Griffithsia corallina	193	Laminaria digitata	1
—— Devoniensis	192	—— fascia	1
—— equisetifolia	191	—— longicruris	1
—— secundiflora	194	—— Phyllitis (Pl. XIX. f. 76)	1
—— setacea	195	—— saccharina	1
—— simplicifilum	192	Laurencia cæspitosa	2
Gymnogongrus Griffithsiæ	220	—— dasyphylla	2
—— plicatus	220	—— obtusa	2
Halidrys siliquosa		—— pinnatifida	
(Plate I. fig. 2)	104	(Plate XV. fig. 58)	2
Haliseris polypodioides		—— tenuissima	2
(Plate III. fig. 10)	137	Leathesia Berkeleyi	1
Halymenia ligulata		—— tuberiformis	1
(Plate XIII. fig. 52)	227	Lichina confinis	1
Hildenbrandtia rubra	226	—— pygmæa	1
Himanthalia lorea		Licmophora flabellata	3
(Plate IV. fig. 13)	114	—— splendida	3
Hypnea purpurascens		Litosiphon pusillus	1
(Plate X. fig. 37)	237	—— Laminariæ	1
Iridæa edulis (Plate IX. f. 35)	216	Lyngbya Carmichaelii	3
Isthmia obliqua	334	Mesogloia Griffithsiana	1
Jania rubens	278	—— vermicularis	1
Kalymenia Dubyi	227	—— virescens (Pl. V. f. 20)	1
—— reniformis	226	Melobesia agariciformis	2
Laminaria bulbosa	122	—— fasciculata	2
—— Cloustoni	127	—— polymorpha	2

	Page.		P
Microcladia glandulosa		Plocamium coccineum	
(Plate VIII. fig. 32)	203	(Plate XII. fig. 46)	
Myrionema clavatum	166	Polyides rotundus	
—— Leclancherii	166	Polysiphonia atro-rubescens	
—— punctiforme	166	—— Brodiæi (Pl. XV. f. 60)	
—— strangulans	165	—— byssoides	
Myriotrichia clavæformis		—— elongata	
(Plate III. fig. 11)	164	—— elongella	
—— filiformis	164	(Plate XIV. fig. 55)	
Naccaria Wigghii	213	—— fastigiata	
Nemalion multifidum	209	—— fibrata	
—— purpureum	209	—— fibrillosa	
Nitophyllum Bonnemaisoni	264	—— formosa	
—— Gmelini	264	—— furcellata	
—— Hilliæ	263	—— nigrescens	
—— laceratum	263	—— obscura	
—— punctatum		—— parasitica	
(Plate XIII. fig. 51)	261	(Plate XII. fig. 47)	
—— versicolor	265	—— pulvinata	
Odonthalia dentata	284	—— spinulosa	
Padina pavonia (Pl. XIX. f. 74)	138	—— subulifera	
Peyssonelia Dubyi	225	—— urceolata	
Phyllophora Brodiæi	223	—— violacea	
—— membranifolius	225	Porphyra laciniata	
—— palmettoides	225	(Plate XIX. fig. 75)	
—— rubens (Plate IX. f. 34)	223	—— miniata	

INDEX.

	Page.
Porphyra vulgaris	322
Ptilota plumosa (Plate VIII. fig. 30)	204
—— sericea (Pl. VII. fig. 28)	205
Punctaria latifolia (Plate IV. fig. 16)	145
—— plantaginea	145
—— tenuissima	146
Pycnophycus tuberculatus	107
Ralfsia verrucosa	168
Rhizoclonium riparium	318
Rhodomela lycopodioides	286
—— subfusca	286
Rhodymenia bifida	243
—— ciliata	245
—— cristata	245
—— jubata (Pl. XI. fig. 43)	245
—— laciniata (Pl. IX. fig. 36)	243
—— palmata	246
—— Palmetta	244
—— sobolifera	247
Rivularia atra	332
Rytiphlæa complanata	290
—— fruticulosa	290
—— pinnastroides	289
—— thuyoides (Pl. XV. f. 59)	290

	Page.
Sargassum bacciferum (Plate III. fig. 12)	96
—— vulgare	96
Schizonema helminthosum	338
—— obtusum	338
—— implicatum	339
Seirospora Griffithsiana (Plate VI. fig. 24)	187
Sphacelaria cirrhosa	158
—— filicina	155
—— fusca	158
—— plumosa	157
—— racemosa	158
—— radicans	158
—— scoparia (Pl. IV. fig. 15)	156
—— sertularia	156
Sphærococcus coronopifolius (Plate XII. fig. 48)	241
Sporochnus pedunculatus (Plate V. fig. 17)	132
Spyridia filamentosa	196
Stenogramme interrupta	248
Stilophora Lyngbyæi	143
—— rhizodes	142
Striaria attenuata (Pl. II. fig. 8)	144
Striatella arcuata	334

	Page.	
Taonia atomaria	142	Ulva Linza (Plate XVI. fig. 61)
Ulva Lactuca	330	Vaucheria submarina
—— latissima		Wrangelia multifida
(Plate XIX. fig. 73)	329	Zonaria parvula

Fresh-Water Algæ.

	Page.	
Batrachospermum atrum	351	Draparnaldia nana (Pl. XX. f. 79)
—— moniliforme		Hyalotheca dissiliens
(Plate XX. fig. 77)	350	Micrasterias denticulata
—— stagnale	352	(Plate XX. fig. 80)
—— vagum	352	Oscillatoria limosa
Cladophora glomerata		—— mucosa
(Plate XX. fig. 78)	344	—— thermalis
Draparnaldia elongata	358	Vaucheria dichotoma
—— glomerata	356	Zygnema quininum

Printed by Reeve and Nichols, 5, Heathcock Court, Strand.

LIST OF WORKS

PRINCIPALLY ON

NATURAL AND PHYSICAL SCIENCE,

PUBLISHED BY

REEVE AND CO.,

5, HENRIETTA STREET, COVENT GARDEN.

---

1. WESTERN HIMALAYA AND TIBET; the Narrative of a Journey through the Mountains of Northern India, during the Years 1847 and 1848. By THOMAS THOMSON, M.D., Assistant-Surgeon, Bengal Army. In one vol. 8vo, with Maps and Tinted Lithographs. Price 15s.

"To all those who desire to judge scientifically of what is possible in the cultivation of the Indo-Alpine Flora, which is so rapidly enriching our gardens, works of this description have great interest. . . . . Unlike gossiping books of travels, the record of researches such as Dr. Thomson's forms a subject of serious reference, which can only lose its value when men cease to regard physical facts as the foundation of all true knowledge."—*Gardeners' Chronicle.*

2. PARKS AND PLEASURE-GROUNDS; or, Practical Notes on Country Residences, Villas, Public Parks, and Gardens. By CHARLES H. J. SMITH, Landscape Gardener. 12mo, cloth. 6s.

3. POPULAR HISTORY OF BRITISH ZOOPHYTES. By the Rev. Dr. LANDSBOROUGH, A.L.S., Member of the Wernerian Society of Edinburgh. Royal 16mo. With twenty plates. 10s. 6d. coloured.

4. HOOKER'S FLORA OF NEW ZEALAND. To be completed in Five Parts. Coloured plates. 4to. 1l. 11s. 6d.

5. HOOKER'S FLORA OF NEW ZEALAND. To be completed in Five Parts. Plain plates. 4to. 15s.

6. SEEMANN'S BOTANY OF THE VOYAGE OF H.M.S. HERALD. To be completed in Ten Parts. Plates. 4to. 10s. plain.

7. THE ARTIFICIAL PRODUCTION OF FISH. By PISCARIUS. *Second Edition.* Price One Shilling.

"The object of this little book is to make known the means by which fish of all descriptions may be multiplied in rivers to an almost incalculable extent. . . . . This principle of increase Piscarius has carried out by argument and experiment in his little treatise, which, we think, is worthy the attention of the legislator, the country gentleman, and the clergyman; for it shows how an immense addition may be made to the people's food with scarcely any expense."—*Era.*

8. POPULAR SCRIPTURE ZOOLOGY; or, History of the Animals mentioned in the Bible. By MARIA E. CATLOW. Royal 16mo, cloth. With sixteen plates. 10s. 6d. coloured.

"The series of popular books on scientific subjects, published by Messrs. Reeve and Co., has been increased by the addition of a treatise on 'Popular Scripture Zoology,' by Maria E. Catlow, who has already contributed other volumes to the series. It contains a short and clear account of the animals mentioned in the Bible, classed according to their genera, and illustrated by a number of well-executed and characteristic coloured plates. It is a seasonable addition to a very nice set of books."—*Guardian.*

"Miss Catlow's abilities as a naturalist, and her tact in popularizing any subject she undertakes, are too well known to need reiteration on this occasion."—*Notes and Queries.*

9. DROPS OF WATER; their marvellous and beautiful Inhabitants displayed by the Microscope. By AGNES CATLOW. Square 12mo, with plates. 7s. 6d. coloured.

"An elegant little book, both in the getting up and its literature. . . . . . The text is accompanied by coloured plates that exhibit the most remarkable creatures of the watery world."—*Spectator.*

"Of the manner in which this work is executed, we can say that, like Miss Catlow's previous productions on natural history, it displays an accurate acquaintance with the subject, and a keen delight in the contemplation of the objects to which it is devoted. As far as the living beings which inhabit 'Drops of Water' are concerned, we know of no better introduction to the use of the microscope than the present volume."—*Athenæum.*

10. INSECTA BRITANNICA. DIPTERA. By F. WALKER, Esq., F.L.S. Vol. I. Price 25s.

11. POPULAR HISTORY OF MOLLUSCA. By MARY ROBERTS. In one vol., royal 16mo. With twenty plates by Wing. 10s. 6d. coloured.

"The authoress is already favourably known to British naturalists by her 'Conchologist's Companion,' and by other works on Natural History. We expected to find in it a useful and entertaining volume. We have not been disappointed. . . . . . The work is illustrated with eighteen plates, beautifully coloured—in most instances affording a view of the structure of the animal. These drawings are not confined to the species living in shells: the various species of land slugs, and the nudibranchiate mollusca, the slugs of the sea, are all described and figured."—*Athenæum.*

12. POPULAR HISTORY OF BRITISH FERNS. By THOMAS MOORE. Royal 16mo, cloth. With twenty plates by Fitch. 10s. 6d. coloured.

"Mr. Moore's 'Popular History of British Ferns' forms one of the numerous elegant and instructive books by which Messrs. Reeve and Co. have endeavoured to popularize the study of Natural History. In the volume before us, Mr. Moore gives a clear account of the British Ferns, with directions for their cultivation; accompanied by numerous coloured plates neatly illustrated, and preceded by a general introduction on the natural character of this graceful class of plants."—*Spectator.*

13. SANDERS'S PRACTICAL TREATISE ON THE CULTURE OF THE VINE, as well under Glass as in the Open Air. Illustrated with plates. 8vo. 5s.

"Mr. Assheton Smith's place at Tedworth has long possessed a great English reputation for the excellence of its fruit and vegetables; one is continually hearing in society of the extraordinary abundance and perfection of its produce at seasons when common gardens are empty, and the great world seems to have arrived at the conclusion that kitchen gardening and forcing there are nowhere excelled. We have, therefore, examined, with no common interest, the work before us, for it will be strange indeed if a man who can act so skilfully as Mr. Sanders should be unable to offer advice of corresponding value. We have not been disappointed. Mr. Sanders's directions are as plain as words can make them, and, we will add, as judicious as his long experience had led us to expect. After a careful perusal of his little treatise, we find nothing to object to and much to praise."—*Gardeners' Chronicle.*

14. POPULAR MINERALOGY. By HENRY SOWERBY. Royal 16mo. With twenty plates of figures. 10s. 6d. coloured.

"Mr. Sowerby has endeavoured to throw around his subject every attraction. His work is fully and carefully illustrated with coloured plates."—*Spectator.*

15. THE TOURIST'S FLORA. A Descriptive Catalogue of the Flowering Plants and Ferns of the British Islands, France, Germany, Switzerland, and Italy. By JOSEPH WOODS, F.A.S., F.L.S., F.G.S. 8vo. 18s.

"The intention of the present work is to enable the lover of botany to determine the name of any wild plant he may meet with, when journeying in the British Isles, France, Germany, Switzerland, and Italy, thus including in one book the plants of a far larger part of Europe than has been done by any preceding author; for Reichenbach's 'Flora Excursoria' omits Britain, France, and the greater part of Italy . . . . and we are not acquainted with any other work of similar scope. . . . . But we must conclude, and in so doing, beg most strongly to recommend this work to our readers, who when travelling on the Continent will find it invaluable; and if studying plants at home, will obtain from it a clue to much information contained in the Floras of other countries, which might otherwise escape their notice."—*Annals of Natural History.*

16. POPULAR HISTORY OF MAMMALIA. By ADAM WHITE, F.L.S., Assistant in the Zoological Department of the British Museum. With sixteen plates of Quadrupeds, &c., by B. WATERHOUSE HAWKINS, F.L.S. Royal 16mo. 10s. 6d. coloured.

"The present increase of our stores of anecdotal matter respecting every kind of animal has been used with much tact by Mr. White, who has a terse chatty way of putting down his reflections, mingled with easy familiarity, which every one accustomed daily to zoological pursuits is sure to attain. The book is profusely illustrated."—*Atlas.*

"Mr. White has prosecuted natural history in almost all its branches with singular success, and in the beautiful work before us has gone far to raise up young aspirants as eager, if not as accomplished, as himself. No book can better answer its purpose; the descriptions are as bright as the pictures, and the kind-hearted playfulness of the style will make it an especial favourite. Unlike some popular manuals, it is the product of first-rate science."—*English Presbyterian Messenger.*

17. VOICES FROM THE WOODLANDS; or, History of Forest Trees, Lichens, Mosses, and Ferns. By MARY ROBERTS. Royal 16mo. Twenty plates by Fitch. 10s. 6d. coloured.

"The fair authoress of this pretty volume has shown more than the usual good taste of her sex in the selection of her mode of conveying to the young interesting instruction upon pleasing topics. She bids them join in a ramble through the sylvan wilds, and at her command the fragile lichen, the gnarled oak, the towering beech, the graceful chestnut, and the waving poplar, discourse eloquently, and tell their respective histories and uses."—*Britannia.*

18. POPULAR FIELD BOTANY; comprising a familiar and technical description of the Plants most common to the British Isles, adapted to the study of either the Artificial or Natural System. By AGNES CATLOW. Second Edition. Arranged in twelve chapters, each being the Botanical lesson for the month. Royal 16mo. Containing twenty plates. 10s. 6d. coloured.

"The design of this work is to furnish young persons with a Self-instructor in Botany, enabling them with little difficulty to discover the scientific names of the common plants they may find in their country rambles, to which are appended a few facts respecting their uses, habits, &c. The plants are classed in months, the illustrations are nicely coloured, and the book is altogether an elegant as well as useful present."—*Illustrated London News.*

19. PHYCOLOGIA BRITANNICA; or, History of the British Sea-Weeds; containing coloured figures, and descriptions, of all the species of Algæ inhabiting the shores of the British Islands. By WILLIAM HENRY HARVEY, M.D., M.R.I.A., Keeper of the Herbarium of the University of Dublin, and Professor of Botany to the Dublin Society. The price of the work, complete, strongly bound in cloth, is as follows:—

In three vols. royal 8vo, arranged in the order of publication . . . . . . } £7 12 6

In four vols. royal 8vo, arranged systematically according to the Synopsis . . . . } £7 17 6

A few Copies have been beautifully printed on large paper.

"The 'History of British Sea-weeds' we can most faithfully recommend for its scientific, its pictorial, and its popular value; the professed botanist will find it a work of the highest character, whilst those who desire merely to know the names and history of the lovely plants which they gather on the sea-shore, will find in it the faithful portraiture of every one of them."—*Annals and Magazine of Natural History.*

"The drawings are beautifully executed by the author himself on stone, the dissections carefully prepared, and the whole account of the species drawn up in such a way as cannot fail to be instructive, even to those who are well acquainted with the subject. The greater part of our more common Algæ have never been illustrated in a manner agreeable to the present state of Algology."—*Gardeners' Chronicle.*

20. POPULAR HISTORY OF BRITISH SEA-WEEDS. By the Rev. DAVID LANDSBOROUGH, A.L.S., Member of the Wernerian Society of Edinburgh. *Second Edition.* Royal 16mo. With twenty plates by Fitch. 10s. 6d. coloured.

"The book is as well executed as it is well timed. The descriptions are scientific as well as popular, and the plates are clear and explicit. It is a worthy sea-side companion—a hand-book for every resident on the sea-shore."—*Economist.*

21. A REVIEW OF THE FRENCH REVOLUTION OF 1848. By CAPTAIN CHAMIER, R.N. Two vols. 8vo. 21s.

"The most accurate and judicious as well as amusing History of the Revolution we have seen."—*Quarterly Review.*

LIST OF WORKS. 5

22. TRAVELS IN THE INTERIOR OF BRAZIL, principally through the Northern Provinces and the Gold and Diamond Districts, during the years 1836–41. By GEORGE GARDNER, M.D., F.L.S. *Second and Cheaper Edition.* 8vo. Plate and Map. Price 12s.; bound, 18s.

"When camping out on the mountain-top or in the wilderness; roughing it in his long journey through the interior; observing the very singular mode of life there presented to his notice; describing the curious characters that fell under his observations; the arts or substitutes for arts of the people; and the natural productions of the country;—these travels are full of attraction. The book, like the country it describes, is full of new matter."—*Spectator.*

23. ILLUSTRATIONS OF BRITISH MYCOLOGY; or, Figures and Descriptions of British Funguses. By Mrs. T. J. HUSSEY. Royal 4to. Ninety plates, beautifully coloured. Price 7*l*. 12*s*. 6*d*., cloth.

"This is an elegant and interesting book: it would be an ornament to the drawing-room table; but it must not, therefore, be supposed that the value of the work is not intrinsic, for a great deal of new and valuable matter accompanies the plates, which are not fancy sketches, but so individualized and life-like, that to mistake any species seems impossible. The accessories of each are significant of site, soil, and season of growth, so that the botanist may study with advantage what the artist may inspect with admiration."—*Morning Post.*

24. ILLUSTRATIONS OF BRITISH MYCOLOGY; containing Figures and Descriptions of the Funguses of interest and novelty indigenous to Britain. *Second Series.* By Mrs. HUSSEY. In Monthly Parts, price 5*s*. To be completed in twenty Parts.

25. THE ESCULENT FUNGUSES OF ENGLAND. By the Rev. Dr. BADHAM. Super-royal 8vo. Price 21*s*., coloured plates.

"Such a work was a desideratum in this country, and it has been well supplied by Dr. Badham; with his beautiful drawings of the various edible fungi in his hand the collector can scarcely make a mistake. The majority of those which grow in our meadows, and in the decaying wood of our orchards and forests, are unfit for food; and the value of Dr. Badham's book consists in the fact, that it enables us to distinguish from these such as may be eaten with impunity."—*Athenæum.*

26. CURTIS'S BRITISH ENTOMOLOGY. By JOHN CURTIS, F.L.S. Sixteen vols. royal 8vo. 770 copper-plates, beautifully coloured. Price £21. (Published at £43 16*s*.)

27. THE VICTORIA REGIA. By Sir W. J. HOOKER, F.R.S. In elephant folio. Beautifully illustrated by W. Fitch. Reduced to 21*s*.

28. THE RHODODENDRONS OF SIKKIM-HIMALAYA. *First Series.* With coloured drawings and descriptions made on the spot. By J. D. HOOKER, M.D., F.R.S. Edited by Sir W. J. HOOKER, D.C.L., F.R.S. *Second Edition.* In handsome imperial folio, with ten plates. Price 21*s*. coloured.

"In this work we have the first results of Dr. Hooker's botanical mission to India. The announcement is calculated to startle some of our readers when they know that it was only last January twelvemonths that the Doctor arrived in Calcutta. That he should have ascended the Himalaya, discovered a number of plants, and that they should be published in England in an almost UNEQUALLED STYLE OF MAGNIFICENT ILLUSTRATION, in less than eighteen months—is one of the marvels of our time."—*Athenæum.*

29. THE RHODODENDRONS OF SIKKIM-HIMALAYA. *Second Series.* By JOSEPH DALTON HOOKER, M.D., F.R.S. Edited by Sir W. J. HOOKER, M.D., F.R.S. In handsome imperial folio, with ten plates. Price 25s. coloured.

30. THE RHODODENDRONS OF SIKKIM-HIMALAYA. *Third and concluding Series.* By JOSEPH DALTON HOOKER, M.D., F.R.S. Edited by Sir W. J. HOOKER, M.D., F.R.S. In handsome imperial folio, with ten plates. Price 25s. coloured.

31. POPULAR BRITISH ORNITHOLOGY; comprising a familiar and technical description of the Birds of the British Isles. By P. H. GOSSE, Author of 'The Ocean,' 'The Birds of Jamaica,' &c. In twelve chapters, each being the Ornithological lesson for the month. In one vol. royal 16mo, with twenty plates of figures. Price 10s. 6d. coloured.

" To render the subject of ornithology clear, and its study attractive, has been the great aim of the author of this beautiful little volume. . . It is embellished by upwards of seventy plates of British birds beautifully coloured."—*Morning Herald.*

32. POPULAR BRITISH ENTOMOLOGY; comprising a familiar and technical description of the Insects most common to the British Isles. By MARIA E. CATLOW. In twelve chapters, each being the Entomological lesson for the month. In one vol. royal 16mo, with sixteen plates of figures. Price 10s. 6d. coloured.

" Judiciously executed, with excellent figures of the commoner species, for the use of young beginners."—*Annual Address of the President of the Entomological Society.*

" Miss Catlow's ' Popular British Entomology ' contains an introductory chapter or two on classification, which are followed by brief generic and specific descriptions in English of above 200 of the commoner British species, together with accurate figures of about 70 of those described; and will be quite a treasure to any one just commencing the study of this fascinating science."—*Westminster and Foreign Quarterly Review.*

33. THE DODO AND ITS KINDRED; or, the History, Affinities, and Osteology of the DODO, SOLITAIRE, and other extinct birds of the Islands Mauritius, Rodriguez, and Bourbon. By H. E. STRICKLAND, Esq., M.A., F.R.G.S., F.G.S.; and A. G. MELVILLE, M.D., M.R.C.S. One vol. royal quarto, with eighteen plates and numerous wood illustrations. Price 21s.

" The labour expended on this book, and the beautiful manner in which it is got up, render it a work of great interest to the naturalist. . . . . It is a model of how such subjects should be treated. We know of few more elaborate and careful pieces of comparative anatomy than is given of the head and foot by Dr. Melville. The dissection is accompanied by lithographic plates, creditable alike to the artist and the printer."—*Athenæum.*

34. A CENTURY OF ORCHIDACEOUS PLANTS, the Plates selected from the Botanical Magazine. The descriptions re-written by Sir WILLIAM JACKSON HOOKER, F.R.S., Director of the Royal Gardens of Kew; with Introduction and instructions for their culture by JOHN CHARLES LYONS, Esq. One hundred coloured plates, royal quarto. Price *Five Guineas.*

"In the exquisite illustrations to this splendid volume, full justice has been rendered to the oddly formed and often brilliantly coloured flowers of this curious and interesting tribe of plants."—*Westminster and Foreign Quarterly Review.*

35. CONCHOLOGIA SYSTEMATICA; or, Complete System of Conchology. 300 plates of upwards of 1,500 figures of Shells. By LOVELL REEVE, F.L.S. Two vols. 4to, cloth. Price 10*l.* coloured; 6*l.* plain.

"The text is both interesting and instructive; many of the plates have appeared before in Mr. Sowerby's works; but from the great expense of collecting them, and the miscellaneous manner of their publication, many persons will no doubt gladly avail themselves of this select and classified portion, which also contains many original figures."—*Athenæum.*

36. FLORA ANTARCTICA; or, Botany of the Antarctic Voyage. By JOSEPH DALTON HOOKER, M.D., R.N., F.R.S., &c. Two vols. royal 4to, 200 plates. Price 10*l.* 15*s.* coloured; 7*l.* 10*s.* plain.

37. CRYPTOGAMIA ANTARCTICA; or, Cryptogamic Botany of the Antarctic Voyage. By JOSEPH DALTON HOOKER, F.R.S., &c. Royal 4to. Price 4*l.* 4*s.* coloured; 2*l.* 17*s.* plain.

38. THE BRITISH DESMIDIEÆ; or, Fresh-Water Algæ. By JOHN RALFS, M.R.C.S. Price 36*s.* coloured plates.

39. CONCHYLIA DITHYRA INSULARUM BRITANNICARUM. By WILLIAM TURTON, M.D. Reprinted verbatim from the original edition. Large paper, price 2*l.* 10*s.*

40. THE PLANETARY AND STELLAR UNIVERSE. By ROBERT JAMES MANN. Price 5*s.*, cloth.

41. ILLUSTRATIONS of the WISDOM and BENEVOLENCE of the DEITY, as manifested in Nature. By H. EDWARDS, LL.D. Price 2*s.* 6*d.*, cloth.

LIST OF WORKS.

Serials.

42. CURTIS'S BOTANICAL MAGAZINE; by Sir WILLIAM JACKSON HOOKER, F.R.S., V.P.L.S., &c., Director of the Royal Gardens of Kew. In monthly numbers, each containing six plates, price 3s. 6d. coloured.

43. HOOKER'S JOURNAL OF BOTANY, and KEW GARDEN MISCELLANY. Edited by SIR WILLIAM JACKSON HOOKER, F.R.S., &c. In monthly numbers. Price *Two Shillings*.

44. ILLUSTRATIONS OF BRITISH MYCOLOGY; containing Figures and Descriptions of the Funguses of interest and novelty indigenous to Britain. *Second Series*. By Mrs. HUSSEY. In Monthly Parts, price 5s. To be completed in twenty Parts.

45. ICONES PLANTARUM; or, Figures, with brief descriptive characters and remarks, of new and rare Plants. Price 2s. 6d. Part XXVII., completing the volume, now ready, price 2s. 6d.

46. NEREIS AUSTRALIS; or, Illustrations of the Algæ of the Southern Ocean. By Professor HARVEY, M.D., M.R.I.A. To be completed in Four Parts, each containing Twenty-five plates, imp. 8vo, price 1l. 1s. Parts I. and II. recently published, coloured.

47. CONTRIBUTIONS TO ORNITHOLOGY. By SIR WILLIAM JARDINE, Bart. Published in parts. Coloured plates.

48. CONCHOLOGIA ICONICA; or, Figures and Descriptions of the Shells of Molluscous Animals. By LOVELL REEVE, F.L.S. Demy 4to. Monthly. Eight plates. Price 10s. coloured.

49. ELEMENTS OF CONCHOLOGY; or, Introduction to the Natural History of Shells and their molluscous inhabitants. By LOVELL REEVE, F.L.S. Royal 8vo. Price 3s. 6d. coloured.

50. CURTIS'S BRITISH ENTOMOLOGY. Re-issued in monthly parts, each containing four plates and corresponding text. Price 3s. 6d. coloured.

LONDON:

REEVE AND CO., HENRIETTA STREET, COVENT GARDEN.

Printed by John Edward Taylor, Little Queen-street, Lincoln's Inn Fields.